Praise for *Terrestrial Architecture*

Terrestrial Architecture is about architecture that does not exist yet. It is in the realm of visionary architecture. Like a renaissance man, Jack Oliva-Rendler sets out to create an interdisciplinary unifying holistic theory, a possible future, or opportunity for designing the planet. The book presents a new paradigm where infrastructure, formal geometries as dynamic formulations of space, data, and ecosystems are intertwined to design the planet in a harmonic way. It seeks to address the planet's natural and climatic urgency through connections of urban form, architecture, and landscape architecture. The reader is led through an exploratory amalgam of subjects from ecology, computation, data, cybernetics, mapping, mechanisms, instruments, and geometry to generate a new future. How will this new architecture take form? How will it go beyond metaphors? How can this holistic type of architecture emerge? What type of political system could enable this type of all-encompassing architecture? It is still yet to come.

> **Benjamin Pollak**, Architectural Designer

In *Terrestrial Architecture*, Jack Oliva-Rendler elegantly weaves the intricate tapestry of our planet's architectural narrative, highlighting the symbiosis between humanity's built environments and the Earth's organic complexity. This book not only reflects a profound understanding of astrobiology and the paradigm shifts needed to achieve sustainability, but also offers a philosophically rich exploration of our role in shaping and being shaped by the terrestrial landscapes we inhabit. It's a compelling read that challenges us to rethink our relationship with our planet through the lens of architecture in all its dimensions and fractals. Rendler elegantly illustrates how: As Above, So Below.

> **Martin E. Wainstein**, Founder, Open Earth Foundation

Praise Continued

Terrestrial Architecture is a brilliant journey through speculative infrastructures that harmonize ecological and human systems in visually striking and always unexpected ways. This book opens whole new terrains of algorithmic design for sustainability grounded in complex geospatial and metabolic contexts.

> **Dr. Stuart Cowan**, Co-Author *Ecological Design* and Executive Director, Buckminster Fuller Institute

With beautiful prose and images, Jack Oliva-Rendler restores a spiritual dimension to architecture that has long been absent from most intellectual fields. After reading this book, you will feel more connected to the earth, and to the human projects that both enhance and endanger it.

> **Graham Harman**, Distinguished Professor of Philosophy at the Southern California Institute of Architecture

TERRESTRIAL ARCHITECTURE

by

Jack Oliva-Rendler

e7 Architecture Studio Press books may be ordered through booksellers or by contacting:
Jack Oliva-Rendler
https://www.jackrendlercerebrum.com/

Because of the dynamic nature of the Internet, any web addresses or links contained in this book may have changed since publication and may no longer be valid. The views expressed in this work are solely those of the author and do not necessarily reflect the views of the publisher, and the publisher hereby disclaims any responsibility for them.

ISBN: (hardback) 979-8-9899403-0-1
ISBN: (paperback) 979-8-9899403-1-8
ISBN: (ebook) 979-8-9899403-2-5

Book Design: Rick Schank of Purple Couch Creative
Editing: Jill L. Ferguson

Dedicated to The Creator of the Universe,
My Heavenly Lord and Savior Jesus Christ

Table of Contents

Prologue

Living with two parents who are architects I was continuously exposed to radical concepts about architecture. Drawings surrounded me as I grew up in my mother's classroom, where she taught architecture at Los Angeles Trade Technical College. The conversations in the family at dinner and on a regular basis were engagingly complex with recurring themes like BIM, Tensegrity Models, the built environment, geometry, etc. It was all quite mysterious to me, but I remained curious and related to architecture via the forms of geometry that I could play with.

In high school, I took an AP human geography course that spoke about conflicts in Africa being geographically generated, the landscape, the resources, the political boundaries, and the environments that were the subject of conflict. This resonated with me on a deep level when I expanded what I was learning in my AP human geography class to encompass the world. Our environments affect us. What surrounds us in our lives and the habitats we live in affect us deeply—physically, emotionally, mentally, and spiritually. Understanding this led me to study architecture.

In college, I felt like I was decoding the history of my life. I asked myself why my parents were so passionate and why was I so concerned also with the same subject matter? I attended the Southern California Institute of Architecture (SCI-Arc) before I went to the Harvard Graduate School of Design where I received my MARCH degree. In college and graduate school I rigorously sought the definition of environmental architecture, chasing the idea of a visionary future with a burning passion. One project stood out among the rest, leading me forward in my journey, the Biomorphic Biosphere by Glen Small, my father's mentor. Things started to make more and more sense as I studied the Biomorphic Biosphere and deconstructed it from many perspectives. It seemed to reveal answers as to how we can arrive at a built environment that is harmonic with the natural system that has been present on Earth for billions of years.

I've encountered a profound novelty in the wonders of nature and its morphologies as a deep sophistication. My explorations in digital architecture and hand drawing are in communication with my

9

philosophical and theoretical research as a perceptual comprehension of environmental architectures. In this book, then, is a robust collection of images associated with in-depth comprehension of terrestrial natural systems, digital technologies that assist such comprehension, and design techniques that allow us to engage the biospheric reality. I've also decided to include thoughts regarding humanity's relationship with the biosphere as a conscious awareness in a philosophical reflection. The awareness I speak of enables certain technical acts of architecture.

I think the book can be inspiring and useful for both the general public and rigorous academics studying or interested in architecture, urbanism, environmental science, engineering, planning biology, philosophy, mathematics, computer science, art, programming, politics, and likely more. I believe it will be a significant contribution to orient ourselves around a planetary existence of humanity linked by the Internet of Things in the Anthropocene having a role to play our environments at the terrestrial scale. Much of architectural discourses as they exist in institutions are concerned solely with parcels and specific sites, while failing to address a fundamental reality and context of the biosphere and planet as a system with billions of years of evolution to create a hospitable space for biodiversity. The conversations regard buildings or city blocks and are not about environments and the design of environments. Meanwhile, on Earth, cities are increasing occupancy by tens of millions, and we build cities in the short time span of a month. We manipulate resources on the planetary scale in unprecedented proportions, the built-environment being the main contributor to such manipulating of resources.

A fundamental misunderstanding exists of how architecture and cities interact with the natural systems because there is no fundamental philosophy of how natural systems and architecture can co-function. The discipline of design is confused to think parcels and autonomy of buildings are the method of achieving environmental architecture.

A greater totality of the built environment exists beyond a single parcel. Contemporary methods of thinking with computational design tools allow for an extended scope of architecture. Our paradigm requires different scales of thinking that are more inclusive and holistic. Urban

Design and Architecture, the design of built-environments, now face the dawn of an AI revolution and algorithmic sophistication ever more evolutionary as computational abilities expand. Our landscapes are now canvases and our cities more manipulable than ever. The tides are turning during the dawn of global deep digitalization. Architecture and interdisciplinary integration with environmental sciences, combined with new computational tools, now enable a new kind of potential project heuristic that is terrestrial in scale and fidelity. Such projects of new scope and intention would resolve the tension between urban and natural environments. The energy dynamics of the Earth have a rate of cyclical creation to maintain an abundance of life. As we become more familiar with terrestrial realities, the grace at which our natural environment flourishes will become our own true affluence and cultivated wellness.

PART 1

Comprehension of the Natural Systems or Reality

CHAPTER 1

Reflections of Ecology, Design, and the Sacred Interconnectivity of All

Architecture's responsibility, as a design discipline, concerns the Earth and its inhabitants. Projects manage natural systems and craft a relationship between the natural environment and its people. The exchanges between matter and living organisms in the environment may be understood in at least a few ways: economically, energetically, or morphologically. As Pope Francis shares in his encyclical *Laudato Si': On Care for Our Common Home*: "The deterioration of nature is closely connected to the culture which shapes human existence." Economies of growth address certain flows and tendencies of built-environment creation; our sociological systems address the flows of such creations as well. Culture as a means of exchanging ideas, communicating, and sharing in a proliferation of willpower are embodied by designs in morphological acts of architecture. Our philosophies and values assume forms as buildings and infrastructures that orchestrate the movement of people and resources and frame connections between them.

But one of the issues with contemporary architecture is that its focus does not include a certain awareness as to which Earth inhabitants its designs are affecting, and this results in measurable tragic consequences. As Pope Francis writes, "The misuse of creation begins when we no longer recognize any higher instance than ourselves, when we see nothing but ourselves." It is as if large quantities of demographics and biodiversity are neglected by the eye of the collective architectural mind. Damaged environments and impoverished societies within built environments originate in flaws of architectural design and a relationship of design to other social systems. A certain sensibility of the Architect to natural and social systems must be acquired to address the lives of biodiverse ecosystems that sustain our populations and lives of organisms within the built-environment.

A post-human paradigm speaks of an acknowledgement of non-human entities like animals, plants, and geologies as having a degree of agency in our environmental systems. Many contemporary projects

15

try to be human centric or humanist, and therefore, overlook the reality that non-human entities have agency also and are even absolutely critical to interdependent systems that allow for sustaining life. There is an "inseparable bond between a concern for nature, justice for the poor, commitment to society, and interior peace," writes Pope Francis.

An existing culture of consumption and an acceleration of such consumptions is embedded in the fabric of our infrastructures and cities, which are in need of fundamental transformations to achieve sustainable movements of resources and energy. But as Pope Francis points out, "We have not yet managed to adopt a circular model of production capable of preserving resources for present and future generations, while limiting as much as possible the use of non-renewable resources, moderating their consumption, maximizing their efficient use, reusing and recycling them."

Sustainable ways of life exist in a cooperation of culture and the built environment that facilitates ecological relationships. Components of ecology exist in layering systems, assemblies, and compositions. Primary to such constructions is the movement of water as means to "health care, agriculture, and industry," as it says in *Laudato Si'*. Cities are dependent on such systems of hydrology and require maintenance and management. The many applications of hydrology proliferate as a discipline of science and are pivotal in engineering and architectural projects. "In assessing the environmental impact of any project, concern is usually shown for its effects on soil, water and air, yet few careful studies are made of its impact on biodiversity, as if the loss of species or animals and plant groups were of little importance," writes Pope Francis. Water remains a primary component of ecology, but the relationships of the environment proliferate across many dimensions.

As we speak of terrestrial realities it is best to discern between a couple terms. Environment may be conceived of what is pervasively surrounding, while ecology may be conceived as studies that derive the interdependency of agents, actors, elements, and system components within the environment as a relationship of living beings. The capacities of architecture to generate ecosystems proliferates as the ability to sustain life of variant forms. The biological studies of codependent relationships between life forms is of most importance for the well-being of the entirety of the planet. The exploitation of resources around the globe, even in remote locations, is linked to our megalopolises where the resources are being consumed. Resources are extracted from places like the Amazon and brought to the United States for consumption via trade routes. The location where resources are being extracted are intricate ecologies while the locations that the resources are brought to are also precious webs of interconnected lifeforms. Resource extraction, manufacturing processes, consumption locations, and supply chains in general, if not balanced with the ecosystem realities, put major strain on biodiversity globally. These are planetary effects. Oil, wood, precious metals, and much more are extracted globally and shipped to our major cities that function as consumption sites. With these distant precious ecologies so far away and our cities covered in concrete it may be difficult to understand the Earth is an interconnected series of ecologies that all link to each other in natural and beautiful ways. These images below are two projections of the same map that I developed at Metabolic Studio to display how shipping routes, displayed by white wisping lines, are moving materials globally to multiple megacity locations, bringing the resources to the ports of the major cities where they can be distributed on land.

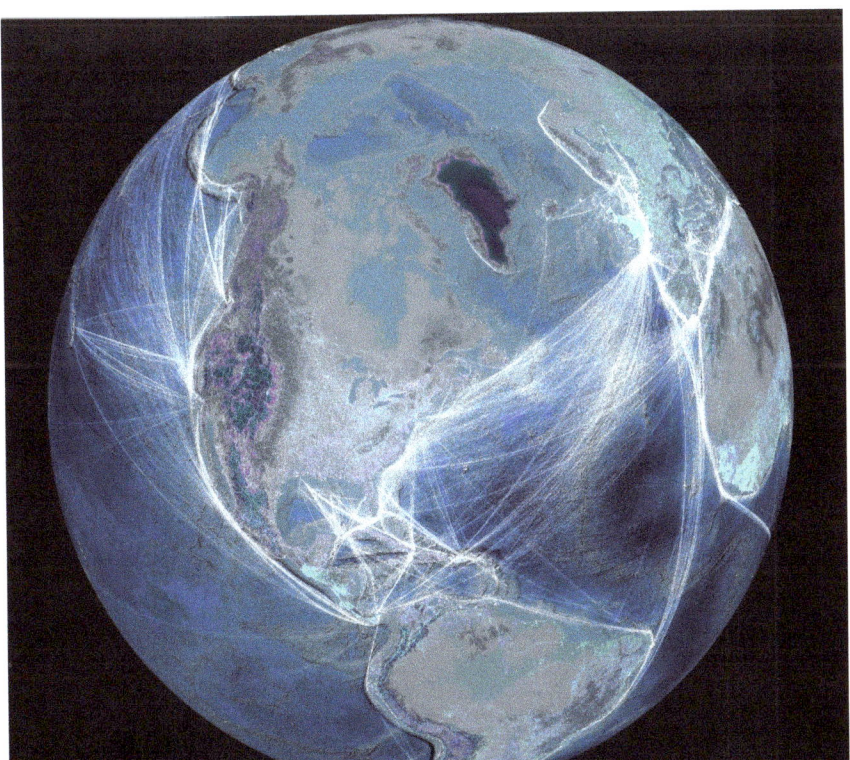

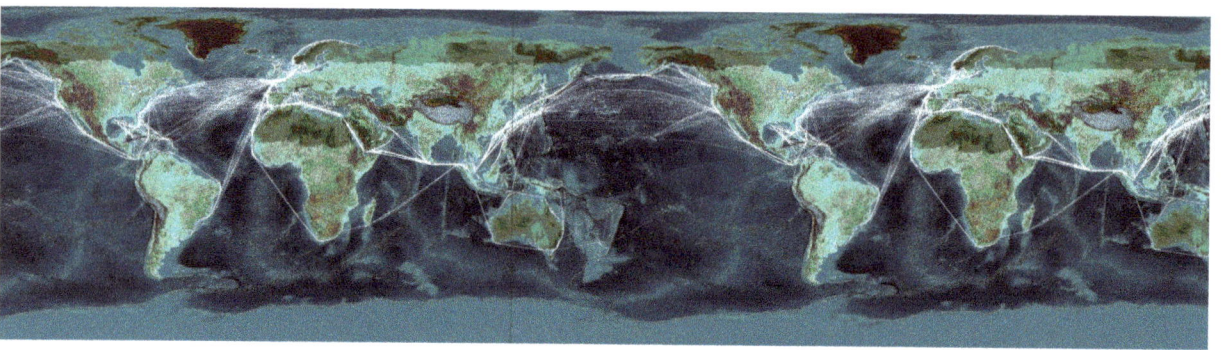

Images I constructed while working with Lauren Bon and the Metabolic Studio

New York City exemplifies humanity's capacities for generating environments that achieve a certain scale of expansion and consumption. Metropolises proliferate in the way a forest would spread and multiply, although the city doesn't produce oxygen or pull carbon out of the atmosphere the way a forest does. Urban megalopolises like New York and the urban sprawl to Boston show the diffusion of architectural structures. An aerial view of Tokyo shows a similar narrative.

As Pope Francis writes:

"Let us mention, for example, those richly biodiverse lungs of our planet which are the Amazon and the Congo basins, or the great aquifers and glaciers. We know how important these are for the entire earth and for the future of humanity. The ecosystems of tropical forests possess an enormously complex biodiversity which is almost impossible to appreciate fully, yet when these forests are burned down or leveled for purposes of cultivation, within the space of a few years countless species are lost and the areas frequently become arid wastelands. A delicate balance has to be maintained when speaking about these places, for we cannot overlook the huge global economic interests which, under the guise of protecting them, can undermine the sovereignty of individual nations. In fact, there are 'proposals to internationalize the Amazon, which only serve the economic interests of transnational corporations'. We cannot fail to praise the commitment of international agencies and civil society organizations which draw public attention to these issues and offer critical cooperation, employing legitimate means of pressure, to ensure that each government carries out its proper and inalienable responsibility to preserve its country's environment and natural resources, without capitulating to spurious local or international interests."

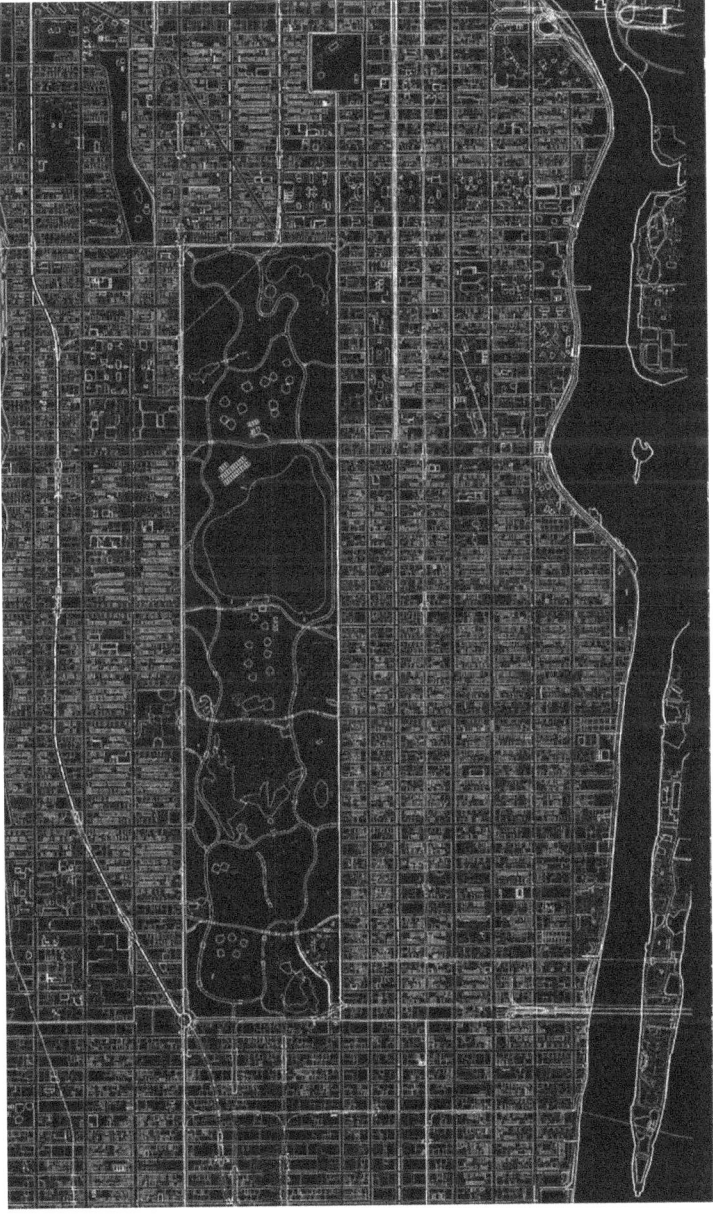

Aerial View of New York City, drawn by me referencing data from the NY GIS database

Such large ecosystems are responsive to human manipulations while their existence and operations provide global sustenance. Ecosystems provide life-giving elements, and only in certain ways, can their usefulness be leveraged to the greatest degree over the duration of generations.

Ecological interdependence is so pervasive that the dependencies on which natural systems operate continuously reactivate themselves in cycles, in processes that have taken place across eons on global scales. As the environment flourishes, symbiotically humanity does as well. The environment flourishes in parallel to our built environment when careful and strategic design approaches the scenario with long-term solutions that engage the complexity on multiple levels of understanding. As the human environment and the natural environment flourish together, the two also deteriorate as one.

As Pope Francis points out, "...we cannot adequately combat environmental degradation unless we attend to causes related to human and social degradation."

Ecological complexity is a design responsibility rooted in the anthropocene, a point in human history as a condition, our position as human beings as having dominion and responsibility over the earth because of our ability to alter its form and chemical composition at the environmental scale. We are at a point where we cannot ignore ecological complexity when it comes to design since we have altered so much of the environment by the way we live and build. Growth in population and consumer culture, or cultures of production, are simultaneous contributors to a societal context of ecological engagement, often in destructive matters. Production processes and waste management become inherently vital to the care for all life depending on their dead-end waste cycles or their adoption of cyclical processes. Business actions around the earth have destructive influence over areas of exchange. Globally resource extraction takes place in certain areas that have natural resources; forests are deforested and mines are stripped of their minerals. The resources are then shipped to manufacturing locations and then to distribution locations. Each location along the supply chain is an augmented factor of the environment.

Rust Belts across the United States are exemplary in showing that a business may leave an area of land and culture unemployed, abandoned, depleted of natural reserves, deforested, and polluted. Our ability to construct and transform the material of the Earth on terrestrial scales is embedded in sovereign structures of government and the private industry's ability to manipulate resources. For example, iPhones are built from minerals in Africa, manufactured in China, and shipped to consumers all over the world. International trade for commercial products is prevalent. A global market dominates most of our megacities as a standard to our rates of production and consumption.

Methods of sustainable development are possible as business ventures and policies—in combination with ecological designs that engage communities.

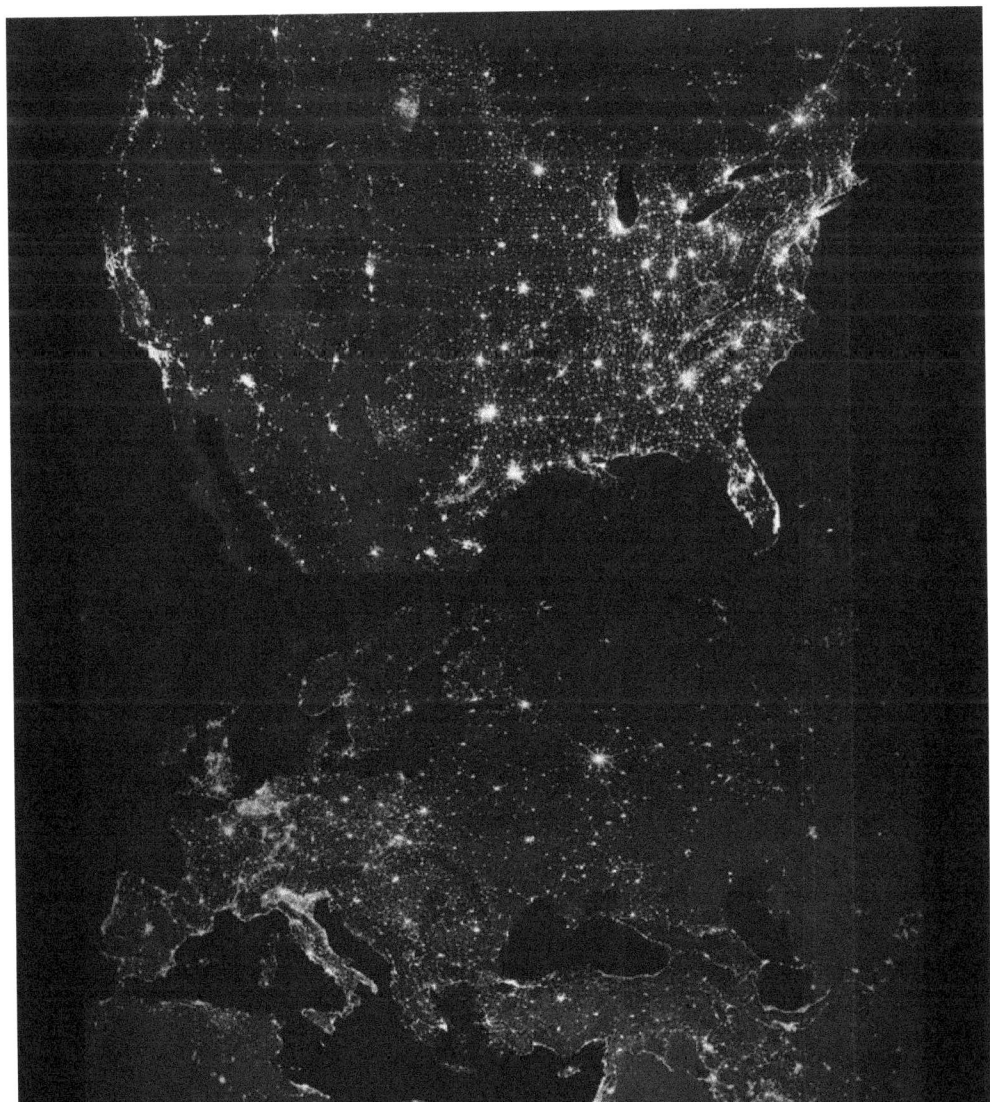

An image of the Earth in the night displays the network of cities distributed across the terrestrial landscape. The cities are linked to a network where their resources are being exchanged and consumed in a simultaneous fashion. This third person perspective of the Earth brings to light the fact that we are in a collective action sequence of inhabiting the Earth and using its resources. We are not autonomous beings, but embedded in a constellation of others with whom we share the planet.

As Pope Francis writes, "God wills the interdependence of creatures. The sun and the moon, the cedar and the little flower, the eagle and the sparrow: the spectacle of their countless diversities and inequalities tells us that no creature is self-sufficient. Creatures exist only in dependence on each other, to complete each other, in the service of each other. This is the basis of our conviction that, as part of the universe, called into being by one Father, all of us are linked by unseen bonds and together form a kind of universal family, a sublime communion which fills us with a sacred, affectionate and humble respect."

Biblical texts address a particular relationship between each human being and the earth. Pope Francis references the texts when recognizing that "they tell us to 'till and keep' the garden of the world (cf. Gen 2:15). 'Tilling' refers to cultivating, plowing or working, while 'keeping' means caring, protecting, overseeing and preserving. This implies a relationship of mutual responsibility between human beings and nature.' The universe did not emerge as the result of arbitrary omnipotence, a show of force or a desire for self-assertion. Creation is of the order of love. God's love is the fundamental moving force in all created things."

Similarly in *GAIA*, scientist James Lovelock writes, "Rene Dubos has powerfully expressed the concept of man as the steward to life on Earth, in symbiosis with it like some grand gardener for all the world." Lovelock believes that the earth is a self-regulating system.

A similar environmental ontology, or a model view understanding of ecology, would be a megastructure as a holistic re-consideration of the relationship between components of the built environment. Glen

Small, architect and co-founder of the Southern California Institute of Architecture, defines megastructures as "a combination of many objects and systems in ways that complement each other and create a diversified but harmonious whole. Nature offers examples of huge ecological megastructures that combine mountains, valleys, forests, streams, animals, etc. Our problem as human beings is to interact with these natural systems in harmonious ways. The ultimate megastructure being the earth as a whole." An interoperability of structures and architecture that self referentially integrated at the global scale would recontextualize the earth as a continuous mega-structural act of ecological consideration.

CHAPTER 2

GAIA and The Biosphere

The term "biosphere" was developed and used by an early 20th century Russian scientist named Vladmir Vernadsky who intended on describing the physics of living matter at the planetary scale. The biosphere came to be understood as a primordial planetary life-support system comprising the peripheral envelope of planet Earth. The term biosphere arose from a proposal around 1875 to unite geophysics with biology. A redescription of the Earth through an understanding of its biological and chemical properties reveals an intricate interconnectivity of living systems where life is maintained through a process of exchanging energy amongst the lifeforms and the Earth minerals. From one perspective, life is expressed through chemistry, through organisms and their relationship to their surroundings as geological and biological forces.

Through consideration of our built environment's capacities for infrastructural augmentations, we can become aware of our reliability and interconnectivity with the larger biosphere systems. Connection to such natural systems is a functional benefit engineered to provide sustenance for life and simultaneously retains an aesthetic expression that is like art, altering the emotional and mental states of experience. One can also have a profound resonant spiritual experience by feeling connected to the cosmic matrix of life.

In cities that are mostly covered in cement people may not feel as connected to the vast ecology of the biosphere, taking part in photosynthesis, releasing oxygen that keeps us alive. But on some level we know that networks of rivers flow through the earth, minerals are within its skin, chemicals exchange with biology, and the Earth is encompassed and embedded with life. We may consider what infrastructural, architectural, agricultural, engineered, scientific, solutions and responses can be made to sustain life in its most abundant form. In a unified science of design we describe and articulate a living planet as we calibrate our infrastructures and built environments to the biospheric reality of our planet and to the life support systems already embedded in the natural physiognomy.

The description of the living planet may begin with solar radiations and transitional states of energy from light to heat and electricity. Solar radiation activates biological matter as the energy is redistributed throughout the Earth, performing in many ways the functions of life. Vernadsky identifies that, "Life is one of those processes which are found whenever there is an abundant flow of energy."

The biosphere as an accumulation of biomass and organisms that form a film or membrane around the earth function as a transformer of cosmic energy, converting radiation into electrical, chemical, thermal, and other forms of energy. Living matter, as cellular organisms of the Earth, perform the translations of energy, generating chemical compounds through photosynthesis and creating energy in the thermodynamic field of the biosphere.

Living organisms and species—plants, animals, and humans of the biosphere—exist in a relationship of interconnectivity; species are engaged and intertwined with each other and their surrounding environments. Each person has the essential responsibility to be fundamentally involved in maintaining and building the environment. The Earth and its inhabitants are bonded; however, such bonds are not always acknowledged in our daily lives or in major decisions of building infrastructure. Our relationship with the Earth as our dependence for the sustenance of life should inspire a love and connection with each other and our environment. Select thinkers in recent history, when acknowledging the collective entity that Earth is, began to call this entity GAIA.

"We have defined Gaia as a complex entity involving the Earth's biosphere, atmosphere, oceans, and soil; the totality constituting a feedback or cybernetic system which seeks an optimal physical and chemical environment for life on this planet. The maintenance of relatively constant conditions by active control may be conveniently described by the term 'homeostasis'," writes James Lovelock.

The chemical processes of living matter, such as photosynthesis as an energetic activity, take place in a particular environment with qualities of temperature, wetness, and exposure to radiation mediated and created by the organisms as features of their being. This condition of the organism actually differs from its environment, considering the temperature of the human body relative to its surroundings. Stable forms of chemistry within the thermodynamic field of living matter cease to function as the organism faces death. This significant compositional activity of the flow of energy through the organism retains an environmental quality of generating surrounding chemical conditions of air quality, ground chemical composition, and light meditations, among other functions. The functions of the biosphere and its relationship with organisms are as old as planet Earth. Only through recent science have we begun to describe in fidelity the function of the terrestrial systems that we interact with in life and when we build our cities. The systems are billions of years old, acquiring evolutionary functionality with only slight variance across distances of time. Throughout geological epochs terrestrial environments have favored the existence of living matter.

Life is a continual influence on the chemical transformations of the surface of the planet. Only green vegetation, which contains chlorophyll (unlike other living systems) makes direct use of solar radiation. The rest of living matter is connected to green vegetation by a direct and unbreakable link. Green matter itself is morphologically transformed and adapted to the cosmic relationship of solar radiation and terrestrial contextualization. Light, even more than food, is the powerful force that determines growth structure. The organisms in their cellular accumulation and structural composition articulate an ability to absorb energy with surface area and angular specificity oriented towards solar radiation and fibrous branching of root patterns that give both means of access to water and structural integrity to the system. Living matter diffuses over the entire surface of the earth as a continuous megastructure of biological tissues producing specific pressures on surrounding environments, interacting with obstacles, overcoming them, only halting when some external force interrupts the movement.

The structure of the organism as a distribution of energy may be translated into architectural forms or algorithmic geometries, like my friend Morgan Garrard did during our studies at the Southern California Institute of Architecture. Garrard experimented with swarm algorithms that generate tree-like pattern structures, where harmony and growth interdependence were apparent.

I noticed something similar on a trip to Hawai'i. On a journey to Kauai, five years later I was consumed by the 1979 book *Godel Escher Bach* by Douglas Hofstadter that I had brought with me to Hawaii from Harvard. Immersed in the middle of its labyrinthian analogies of algorithms and forms, I encountered one of Earth's most biodiverse ecosystems and the variability of the flowers struck me as a robust geometrical language derived from the process of exchanging energy with the biosphere. I bought an encyclopedia of the flowers found in Hawaii, two volumes titled *Manual of the Flowering Plants of Hawai'i*. The anatomy of the flowers is robust in diversity yet share similarities in how to be structured in relationship to terrain; they reach tall enough and provide surface area for light to be absorbed, as shown in my drawing below.

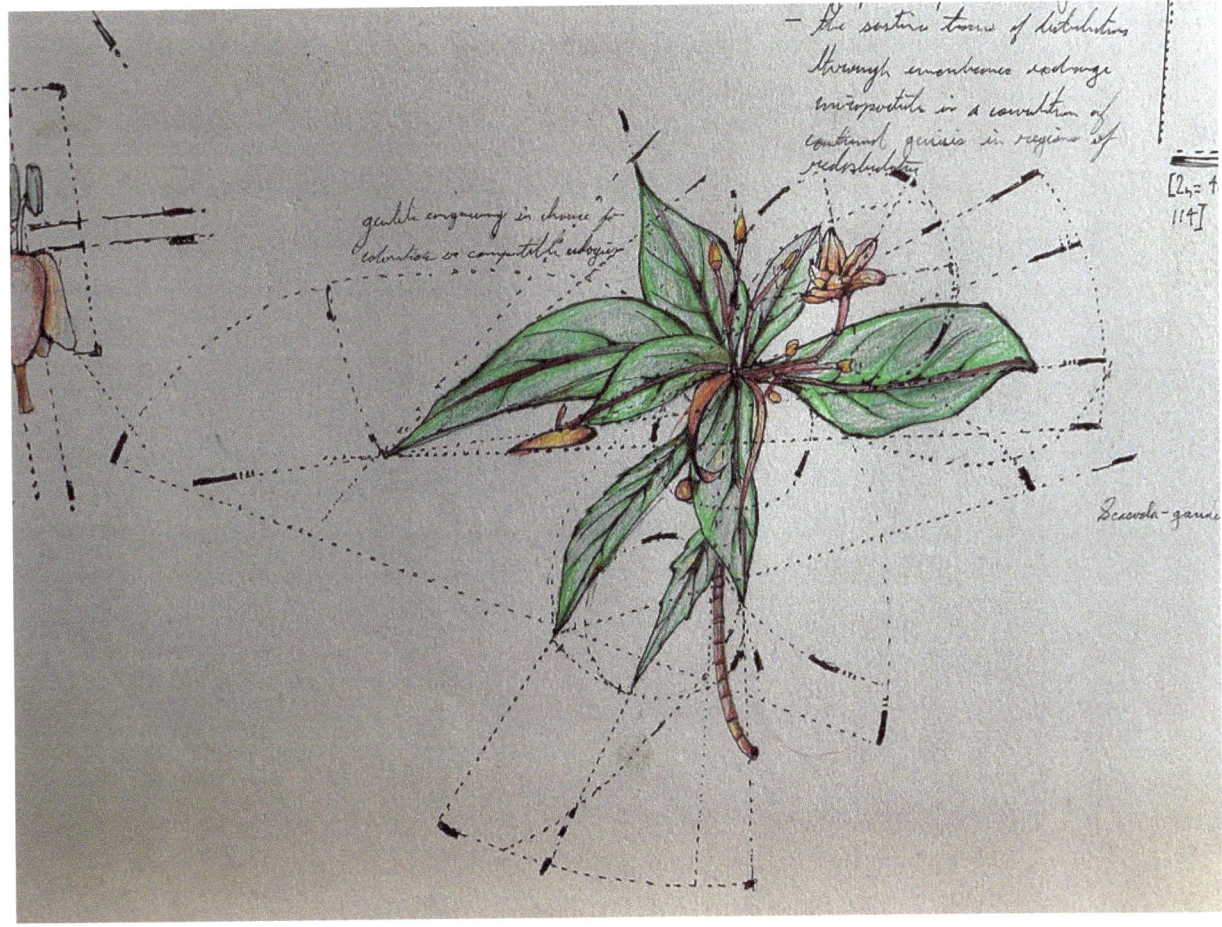

Drawing by Jack Oliva-Rendler

The wide spreading or diffusion of living matter on the planet's surface is caused by the multiplication of organisms as they occupy space. The movement of the multiplication as a proliferation of life in the biosphere is measured by the transformation of chemical elements and the creation of new matter from them. This proliferation of energy of life in the biosphere is referred to as "the geochemical energy of life in the biosphere" by Vernadsky. The movement and multiplication of living matter in the biosphere is occurring through gaseous exchanges between the living matter and its surrounding medium in which it moves. This breathing mechanism is a controlling force that discerns rates and means of multiplication. The movement of air, water, and heat radiation throughout the environment is a morphological correspondence with both the living matter and the built environment matter, the organic matter and the artificial architectural constructions of infrastructure within the built environment.

Another variable that influences rates of multiplication is the living matter's density. The accumulation of these energy systems cooperating is part of the mechanism of the biosphere. The respiratory system provides the fundamental relationship of the organism to the rest of the biosphere. As our bodies have evolved to adapt to the conditions of the Earth so must our architecture be contextualized with a biosphere ecology and a social interconnectivity of humanity.

"The gasses of the biosphere are identical to those created by the gaseous exchange of living organisms. Only the following gasses are found in noticeable quantities in the biosphere, namely oxygen, nitrogen, carbon dioxide, water, hydrogen, methane and ammonia. This cannot be an accident. The free oxygen in the biosphere is created solely by gaseous exchange in green plants, and is the principal source of the free chemical energy of the biosphere. Finally, the quantity of free oxygen in the biosphere, equal to 1.5×10^{21} grams (about 143 million tons) is of the same order as the existing quantity of living matter, independently estimated 10^{20} to 10^{21} grams. Such a close correspondence between terrestrial gasses and life strongly suggests that the breathing of organisms has primary importance in the gaseous system of the biosphere; in other words, it must be a planetary phenomenon," writes Vernadsky. There is a spatial condition that is attributed to respiration, transpiration, and solar absorption simultaneously in the biospheric architectures. The manifold energetic processes are properties of the morphologies that the living systems assume in their form. Both energy transmission methodologies of respiration and solar radiation are properties of the biospheric system and define its structure. Energy transmission in photosynthesis happens through an aggregate of cells as a surface area absorbing solar radiation.

The surface of the leaves absorbing the sun's energy have a microscopic tectonic architecture that has a certain proficiency of energy absorption and exchange within the entirety of the structure. In the ocean, solar radiation penetrates to greater depths and accesses greater volumetric space that contain unicellular algae. This spatial ratio of radiation to living matter is a condition that creates living matter at a comparably faster rate than the plants and trees of land. This includes the phytoplankton, enabling an even greater quantity of solar energy into living mass. Such morphological proliferations are the result of energetic principles. The morphological consequences of volumetric space with permeating light can translate into vast fields where light penetrates continually like forests and the forest floor. Such multi-layered light porosity and depth may be able to be achieved in architecture.

Image I took in the Redwood forests of Santa Cruz, California

Burroughs-Wellcome Dining Area. Image Courtesy of the Paul Rudolph Estate and the Paul Rudolph Institute for Modern Architecture

This image from Paul Rudolph is the multi-layered stacking of space without the loss of natural light. There is also the attribute of extended depth and perspective that gives the presence of a field, a vast expansion of the architectural system resembling the diffusion of the organic system as in the forest.

Any space frame model either by Konrad Wachsmann or Buckminster Fuller resembles a tree canopy with a surface area above and an enclosed environment below. The structural system is cellular, similar to the nature of an organism. A space frame structure as a cellular and modular system may function as a prototype for how we think about environmental architectures in terms of the way components are aggregated.

As we construct our infrastructures and accumulate mass we compose a larger network of energy exchange in the form of heat, light, electricity, water, and respiratory air. The future of architecture will include a familiarity with the ecological processes to then participate with and leverage the energy systems available—to the greater health and well-being of the biodiversity in the planet.

Part of this familiarity includes understanding that organisms in their internal processes pass chemicals through transformative phases and states before the energy or matter is lost to the organism. The multicellular system makes choices as to what is to be discarded or kept. Chemical compositions within the biosphere have locality and contextuality; thermodynamic equilibrium resolves existing contextual scenarios. Like words within a sentence or variables within an equation, the dynamics of respective components give activity integrity and necessity to contextual systems. Contextual conditions, environments of chemical compounds, and environmental constructions provide conditions for life. An inherent momentum is automated in the biosphere, but humanity through our constructions has the capacities to interact and operate on and with such systems through means of agriculture, irrigation, cross ventilation systems, solar energy absorption, mass gardening, natural reserves, and more.

Essential to the comprehension and cooperation with the natural system is the ability to sense and see its performances. A human ability to perceive the environment and understand its properties and then reflect an act of transformation may be conceptualized as an ongoing process called a feedback loop. American mathematician Norbert Wiener created the concept of a cybernetic feedback loop as he studied an animal organism and a machine as a system with sensory awareness and responsive activity. Coming to an understanding of the natural systems happens as we identify relationships between organisms and their environments as systematic processes. The translation of ecological and environmental systems into information systems in documents such as atlases and indexes of species brings us to a greater familiarity with natural systems. Embedded within ecological processes are rules and tendencies of operation. Certain chemical combinations and conditions create life and sustenance for it. The conditions manifest at multiple scales, such as entire river watersheds that are thousands of miles wide and long and in specific ponds that may be a circle with a 10 foot diameter. The conditions for life are assembled in a cooperation of components and elements that send information to each other via the form of energy.

As Lovelock writes in *GAIA*, "Information is an inherent and essential part of control systems in another sense, that of memory. They must have the capacity to store, recall, and compare information at any time, so that they may correct errors and never lose sight of their goal. Finally, whether we are considering a simple electric oven, a chain of retail shops monitored by a computer, a sleeping cat, an ecosystem, or Gaia herself, so long as we are considering something which is adaptive, capable of harvesting information and of storing experience and knowledge, then its study is a matter of cybernetics and what is studied can be called a 'system'."

James Lovelock as well as Vernadsky both identify a feedback between living systems, organisms and environments that create conditions for life, Lovelock draws an analogy of environmental conditions and feedback loops that maintain as a planetary process of homeostasis. The process of homeostasis within the organism as maintaining an equilibrium of energy conditions, like temperature and other chemical balances, is used to describe a planetary maintenance of equilibrium.

"The most important property of Gaia is the tendency to keep constant conditions for all Terrestrial Life. Provided that we have not seriously interfered with her state of homeostasis, this tendency should be as predominant now as it was before man's arrival on the scene," writes Lovelock.

CHAPTER 3

Geochemistry

In continued reflections from Vernadsky we understand an intricate functioning of ecology and the environment through both its biochemistry and its geochemistry, which offers a complete picture of the scale and totality of the biosphere. This way of thinking assumes common traits and tendencies in behavior and action across environments, bioregions, and species as a certain nature of the totality. Vernadsky writes, "A prime example of an empirical generalization is the periodic table of elements. When enough was known about the valences and atomic masses of enough elements, it became apparent to my professor, Mendelyev, that they obeyed periodic law and this law was so powerful that it predicted the existence of many more elements."

All living creatures are embedded in the environment and depend completely on its energy substrates for all forms of nutrition and sustenance for life, including the oxygen in the atmosphere and the chemical compounds of food that come from the Earth and its food chains—originating in energy from the sun. Chemistry is throughout and within the environment and living organisms that bring functionality to both. Chemical elements are not distributed chaotically throughout space and time. At any celestial or terrestrial scale the elements adhere to the structure of their atoms as a system, which informs their composition. Atomic compositions inform the formation of stars, galaxies, nebulae, geologies, organisms, and environments in a fundamental quality of all matter of the universe.

In *Geochemistry and the Biosphere*, Vernadsky writes, "One hundred fifty years after Huygens, the Englishman W. Hyggins, through scientific experiment and observation by spectrum analysis, proved the identity of chemical elements (atoms) of the stellar worlds based on terrestrial manifestations. The present-day creative explosion of ideas has not shattered this essential principle. He expressed it in the new concept of the identity of the basic elements (electrons, neutrons, protons and the newly discovered positive electrons or 'positrons') which make up atoms or chemical elements, and also in that of the genetic, though complicated connection existing between the atoms of different structures."

The atomic foundation of matter reveals discernable tendencies within the elements to react in guaranteed ways revealing fundamental consistencies of energy transition correlated to the formation and transformation of elements. Such consistencies are identifiable processes within the Earth with which architecture and urban environments become cofunctional. One discernable behavior is a cyclicality of the elements to transform in reversible processes.

"The third group of cyclical or organogenic elements is the largest in mass. It includes the greatest number of chemical elements and makes up almost the entire Earth's crust. It is characterized by numerous reversible chemical processes. The geochemical history of these elements may be expressed by cycles. Each element gives compounds characteristic of a certain geosphere; these compounds are constantly being renewed. After more or less long and complicated changes, an element returns to its initial compound and begins a new cycle. This character of terrestrial chemical reactions was noticed for oxygen in the second half of the eighteenth century; the great scientists of that time, who had discovered the terrestrial gasses and their properties, foresaw these characteristic chemical cycles," writes Vernadsky.

Cyclical transformations of elements as reversible processes— an example being the movement of oxygen through the biosphere and the releasing of carbon dioxide from biologies— are movements of chemical energy through a system of exchange. Behaviors and actions of the chemical elements are in connected relationships and carry out transformative work on the terrestrial environment over time. The transformations of terrestrial systems that occur because of biochemical and geochemical activity are most often migrations, movements, and shifts. Such shifts occur through material transitions of liquids, gasses, and solid bodies in changing temperatures. Processes of breathing, circulating fluids, nutrition, and the metabolism of organisms may all be considered migrations. The transition of energy through biology and geology is a way of achieving atomic equilibria in static and dynamic systems. Rules govern

38   composition, but the rules are often bent or broken to accommodate new confluences of organism and environment.

The forms of living bodies are more significant than the substances that pass through them. Their structure and inherent functions through organic connections and anatomic configurations will determine the functionality of the energy that passes through them; however, the system is constantly penetrated by exterior molecules. As long as this transitioning of energy through the organic system is maintained, the organism lives. The chemicals that move through the organism are a direct result of the relationship that has developed between the environment and the organism. The principle that the form of the organism and its organs controls the elements that penetrate it can be extrapolated as an architectural principle. The membranes that envelop an organism may be analogous to the permeable exterior surfaces of infrastructures and buildings, and cities that orchestrate the movement of energy through themselves. The structures that migrate the chemical energies are more influential than the chemicals themselves as catalysts to processes.

In our built environments and infrastructures that organize the flow of people and resources, we are affirming the transition of chemical processes through the physiognomy of our constructions. As the body discerns the processing of nutrients, the city—in its infrastructure or edifice—discerns the processing of terrestrial resources. As cities and built environments increase in complexity in response to the bandwidth of greater population and demographic dynamics, it is not farfetched to imagine that its chemical compositions are parallel to the biosphere and its organisms in terms of complexity and influence on the environment. Certain elements can be identified as being influential on the environment.

For example, oxygen, as a massive component of the atmosphere and an essential ingredient to the life of the organism, influences the physiological transformation of the environment. The chemical constructions that are built by life forms in relation to their environment are architectures that achieve greater scale and influence on surroundings as consumptions and circulations amass. Verndasky uses the examples of corals and lime algae construction over thousands of square kilometers as examples.

39

In life, organisms and the environment are only formed in response to chemical circumstances due to the relationship between said organisms and their environments. One can imagine combinations of geological circumstances with hydrological contexts like the delta of the river Nile cultivating a culture and economy. The environment can be understood in its ability to nourish, create, and move resources into relationships, into populated landscapes and aspects of ecology.

Living matter or organisms in relation to their environment are systems dispersing their translations of energy across the surface of the planet. The chemical physiognomy of the planet is fundamentally defined by the anatomical functions of organisms that through chemical processes reconfigure molecular structures in environments in accumulations and massive scales. Organisms move oxygen and carbon throughout the environment with consequences of thermodynamics and equilibrium and nutrition for biologically-diverse ecosystems. Organisms are composed of water and oil affecting the hydrological absorptions and dispersions on the surface and deep within the planet. The dispersion and embedded presence of oil and coal in the earth is sourced through the organic processes of generations of organisms decomposing into the earth—it is a finite resource even in the contextual abundance of the planet.

"The formation of oil is an extremely important manifestation of the process in which solar energy is transferred to deeper layers of the planet through living matter," writes Vernadsky. This is a multi-billion-year-old system of energy translation in the biosphere. Yet humans destroyed and are destroying these precious products of oil and coal, abusing the substance with usage that exceeds necessity and creating carbon emissions that warm the planet, a characteristic of ignorance—without thinking about the future. Perhaps these reserves of solar energy deep within the Earth that manifest as oil should be considered with a different reverence than today, for sustainable particular uses that consider the longevity of the planet, and not for the exploitation of quarterly gains. The contemporary use of oil exemplifies how we are operating in a different spatio-temporal dimension than the biosphere. This ratio of consumption and production does not exemplify a reverence for the cyclical energetic processes of the

Earth that not only sustain life for long periods of time but also create an abundance of life. "The formation of coal is connected with marshes, with large accumulations of plants characteristic of countries with moist climates, for example in the mouths and deltas of big rivers, in the plains of their basins, on the coasts of continents and islands, and in the lowlands of tidal areas. All these are accumulations of life in which a large mass of organic matter is present in a state of slow decomposition. Possibly they are the largest terrestrial accumulation of life we know," writes Vernadsky.

We can identify robust chemical expressions of life from the presence of organic processes on our planet. The translation of energy as we harvest the earth is location-based and is directly related to our industrial manufacturing processes and our built environments. Instead of a purely extractive model of consumption we should move towards imagining productive co-operations with existing abundant technologies of organic production, which would include acknowledging the cyclical nature of biospheric chemical processes.

Perhaps nature is a technology. The root word *Techne* means to know how. Nature as a technology would mean that it has, in a sense, learned to have certain traits in its being to process energy. Axiomatically organisms are multiplying and proliferating in context to their environment using available energy conduits in the Earth and thereby generating new energy that is available. The biosphere is so robust that the atmosphere itself is composed of elements, and the biosphere and the atmosphere complete each other in cycles of chemical exchange. Or, as Charles Darwin wrote, all locations of the world are supportive of life.

The shapes and forms of the organism defined by its environment and the contents of environments as morphologies defined by the organism as a cyclical feedback loop is fundamental to science and is overlooked in architecture. Transmutation and proliferation of biology defines the physiognomy of the planet in the anatomical constructions of biology that diffuse across the surface of the earth, manipulating energy in the temporal process of its architecture. As Vernadsky states, "The most distinct empirical generalization dominating all our ideas about the Earth's realm is the incessant and regular change of the morphological structure of living matter—the evolution of species. The idea of the evolution of the organic realm seems one of the greatest achievements of the human mind in the last century. It is obvious that morphological change must be closely connected with chemical change, as the form of an organism, like the form of any physical body, is determined first of all by its internal chemical structure." The physical structure of living things as conduits for energy is the way in which Vernadsky derives general principles regarding the biosphere. The structures, shapes, and forms that living matter and environments assume define the dynamics of energy in the environment.

As Vernadsky emphasizes the importance of the morphological structure of living matter, he reveals a future body of work that lies in the field of architecture. "The organisms' work in transforming solar energy into terrestrial chemical energy is subject to immutable mathematical laws. Everything is calculated and proportioned to the precision, rhythm, and measure which we see in the harmonious movements of a heavenly body which we are beginning to see in the constituents of matter and energy, such as atoms." As Vernadksy speaks of the organism transmitting energy using mathematical laws, how it is all proportioned, rhythmic, in the measurement of harmonious movements these terms deeply resemble the sentiments of contemporary parametric, digital architecture, where forms of architecture are calibrated towards specific variables that drive the form.

Meditations on the nature of form and morphology proliferate in this book as methods of composing dynamic integrated systems of geometry can be integrated with the energy processes of biology. As the organisms mitigate the relationship between solar radiation and the planet, the physiognomy of the planet is defined. Organic material is the arbiter for solar energy on the planet, translating and transitioning the energy states into a myriad of functionalities that define our environments, geologies, and even atmospheres. There is an architecture to organisms and an architecture to environments and a feedback loop between the two. Here architecture is thought of as forms and shapes of components and

42   systems within the organism and in the environment. There is a method to which the pieces of the environment and the organism come together in such a way that they are synergistically dependent on one and other.

Vernadksy writes, "The organism's proliferating apparatus is a specific mechanism for the dispersion of the geochemical energy of life, regulating elements' migration in the biosphere and thus the entire Earth's crust. The speed of dispersion in organisms that are the most adapted for this purpose reaches their physical limit. Thus, living matter becomes the regulator of the biosphere's active energy. It distributes the latter more or less uniformly over the Earth's surface. Therefore, the surface layer of the planet becomes, through living matter, a field of manifestation of kinetic and chemical energy."

Chemical energy in the organism originates from the solar radiation of the sun that illuminates the Earth. The sun fills the Earth with energy through radiation and the radiation becomes omnipresent and forms of light, heat (thermal) mechanical, or electricity and can transform material surroundings. Using photosynthesis, solar rays create new compounds and free energy throughout the thermodynamic biosphere.

CHAPTER 4

Philosophy Biology and Morphology of Nature

O ur philosophical perception of nature influences our daily actions, social interactions, our methods of doing business and creative projects. Pope Leo XIII in his Encyclical Letter mentions the potential of philosophy to influence culture into a way of behaving and perceiving, this can have positive consequences or it can be dangerous. This quote is from the Encyclical Letter as noted by Thomas Aquinas:

> "It is proved by fact and constant experience that the liberal arts have been most flourishing when the honor of philosophy has stood inviolate, and when its judgment has held for wisdom: but that they have lain neglected and almost obliterated when declining philosophy has been enveloped in errors and absurdities. Hence, also, the physical sciences, which are now held in so much repute, and everywhere draw to themselves a singular admiration, because of the many wonderful discoveries made in them, would not only take no harm from a restoration of the philosophy of the ancients, but would derive great protection from it. For the fruitful exercise and increase of these sciences is not enough that we consider facts and contemplate Nature. When the facts are well known we must rise higher, and give our thoughts with great care to understanding the nature of corporeal things, as well as to the investigation of the laws which they obey, and of the principles from which spring their order, their unity in variety, and their common likeness in diversity. It is marvelous what power and light and help are given to these investigations by scholastic philosophy, if it be wisely used."

Even if the participant in societal culture, a citizen of the world, is not a philosopher they still affect society through their actions, inevitably subscribing to a perception of their relationship to their surroundings. When our existence plays out on planet Earth our relation to nature becomes very influential on our decision making. When we become familiar with the physical sciences, our understanding of the natural system evolves and simultaneously our practices of living and designing, for example, understanding nutrition and eating healthier and understanding gravity which has led to novel structures, such as the flying buttress and the composition of steel to make Eiffel Tower, etc. Sometimes there is an attempt to domesticate and normalize the earth in a simplified process of commercialization and commodification. The genuine sophistication of the earth's inner workings as an evolution of behaviors and technicalities over billions of years is mostly ignored or over summarized for market purposes while a philosophical process would allow for the more expansive ever-present reality of the Earth to come into greater form.

The center of the contemporary philosophical debate to comprehend nature is the analogy of the machine as there is a history in explaining the functions of nature from the bottom up in terms of its parts, explaining the nature of behaviors and components of nature as a mechanism. The history of science and industrialization, when approaching creative projects or analytical understandings of our economies and environments, is embedded culturally with a tendency to (aggressively) isolate aspects of things with autonomy—objects, species, materials, commodities, societies, people, etc. of our world in crude objectivity—as if to perceive an individual function of our material world without its relationship to other functions. Industrial cities are continually constructed on the basis of production of both machines and organisms, as in automobiles and agriculture, by leveraging what resources are accessible by the business corporation and government. This economic production of nature when optimizing isolated uses and exchange values often ignores ecological interconnectivity or the many dependencies that are shared within an environment, resources and their movements as sources of energy. The long history of attempting to reduce nature into a simplified construct that can be controlled, abused, taken advantage of, or even neglected has resulted in critical side-effects, hence the carbon in the atmosphere, pesticides in the crops, plastic in the ocean, smog in the air ,and degrees of consumption and resource distribution that leaves a large portion of the population on Earth degraded or starving. Autonomy and separation of the natural system into isolated components from itself is a folly. It is a foolish approach, even if in some illusion of a construct it may seem profitable. It is simply a fantasy that will run its course into disaster. We are destined for harmonic synergetic integration in the unified functioning of earth systems.

To comprehend nature in detail by the scientific method of isolating objects with objectivity as in analyzing subatomic particles, atoms, and molecules is valuable only in the ecological context that the reactions and potentials of energy that are discovered are appropriated for projects that take action to ensure wellness and life for all of biodiversity. The discoveries in the technicalities of nature that serve medicine and sustainable engineering are truly beneficial to support the longevity of our precious biodiversity. The leveraging of scientific knowledge in architecture ensures that we can be more responsible agents of service and stewardship to all life on earth. The whole of our environment as a natural system must constantly be referenced, to discern the actual function of each part, to maximize our participation with the complete inertia of energy in which the universe has set its course, sometimes referred as the Dao. The typical objectification of the earth focuses on a limitation of its being, rather than an expansion of its potential. We can participate in the radiant flow of creation or stubbornly refuse and face the wrath of ignorance that exposes us to suffering in states of lacking energy as opposed to an abundant reality of an infinite universe. It's important to contextualize isolated members of scientific objects, integrating them into larger systems catalyzing manipulations of energies of the earth.

Graham Harman, in *Object-Oriented Ontology: A New Theory of Everything,* states one of his criteria for objecthood as "Objects come in just two kinds: real objects exist whether or not they currently affect anything else, while sensual objects exist only in relation to some real object." He elaborates further saying, "Real objects cannot relate to one another directly, but only indirectly, by means of a sensual object." One can imagine in geochemistry the many elements of the earth coming into contact with one another, forming erosions, eruptions, diffusions, transmutations of solar energy, biochemical morphogenetic feedback loops that exert physical effects on surrounding matter, and more. Aristotle would even call the affectual reality that seemed to be beyond physics, metaphysics.

Graham describes,"An object is any thing that cannot be entirely reduced either to the components which it is made or to the effects that it has on other things."

Graham and other contemporary philosophers like Timothy Morton use an object- oriented ontology to discuss larger objects called hyperobjects that are so vast in space and time that they are a mystery. When we discuss entities like the East India Trade Company or A Building composed of many pieces there are just two ways of telling somebody what something is: you can tell them what it is made of, or tell them what it does. Entities and objects like river systems and infrastructures, geological systems and the blurred boundaries of how they define each other simultaneously, as in the case of the Colorado River, make significant case studies for the comprehension of an object. An entity like God is so vast and omnipotent that to escape its all-encompassing presence is impossible. Philosophically it is important we come to an understanding of these grand forces and interact with them in our architecture. The mystery of our existence, our relationship to the Earth and each other depends on this. Bruno Latour identifies a tendency of modernism to classify two main agents at play in the paradigm, nature and culture, however likely the many branches of thought and action that take place segment nature and everything else, as if reality is capable of being separated from itself. Furthermore, as one considers that "nature" or "reality" can be separated into controllable components, we find ourselves in a space of trying to control the earth or even the cosmos. Philosophy is a grounding ethos of action, and it provides a basis to see the interconnectivities of our ecologies.

Forces of nature, forces unchanging, laws of nature, laws of physics, evolution, systems, processes of natural systems, etc. are all complex vast realities to be met with novel, innovative, thoughtful intelligence. That which is changing and that which does not change—how can we measure such things and design intelligently collectively? Here also lies questions of dynamic potentialities and probabilities.

The Earth is likely more abundant than we can imagine; the Earth's life systems may be at risk because of humanity's dysfunctional behaviors, ways of life, and our built environments, which follow false philosophies of lack, scarcity, and autonomy. Even constructs and practices of intellectual thought or design agendas, which centralize the importance and value of ecology, tend to still define the Earth as limited as compared to humanity's power over it. Culture is one level of how behaviors are disseminated, yet through the art of rituals and ceremonies, core beliefs about the human and our relationship to the Earth are illustrated. One does not need to look far to find indigenous cultures who understand their ecologies in intimate ways and who do not seek to dominate them. Embedded in all relationships is a grammatical language Algerian-French philosopher Jacques Derrida describes as the transcription of reality to symbols that carry meaning for people. Many indigenous people see plants, animals, the sky, and mountains as living at their peak with words that carry the meaning that they are alive. Our languages define our perspectives and enable us to act.

In Bruno Latour's book *Facing Gaia: Eight Lectures on the New Climatic Regime,* he writes, "Before looking into what must be done, we must be impelled to action by a particular type of utterance that touches our hearts in order to set us in motion—yes, to move us. Astonishingly, this type of utterance now comes not only from poets, lovers, politicians, and prophets but also geochemists, modelers and geologists.

The utterances of this book explain an interdependence from multiple perspectives and approaches in order to derive overarching and underlying consistencies to the nature of the Earth or its propensities as a systemized entity in which we may exist in better harmony with its dynamics, including each other.

Latour quotes James Lovelock by saying, "Take water. It should have vanished long ago, just as it did on the other planets. Why is it still here, and so much of it? 'The Earth has abundant oceans because it has evolved, not by geophysics and geochemistry alone, but as a system in which the organisms are an integral part.'"

I may have derived an inscribed truth in reality that I've come to understand is the principle of algorithmic morphologies: the language that Earth speaks in its physiognomy in chemistry and biology as architectures is a morphological ontology as systems become cofunctional. The fundamental manipulation of energy in nature happens through morphological phenomena which I consider architecture.

The following explanation of morphological phenomena may differ from an original description of "Morphic Resonance" from Rupert Sheldrake. Still, his terminologies and definitions are useful. His theory proposes that the morphological formation of all matter is affected by all other matter that follows similar formation. The difficulty I attribute to his theory is in the concept of action from a distance. This idea becomes challenging to use as a basis from which to make decisions because matter affecting other matter happens by recording the accumulation of processes in nature so that all past processes become presently causal, affecting the present, perhaps by a memory retention of all matter. Rupert Sheldrake names the communication between matter across time and space as "Morphic Resonance". The resonance of matter across space-time as action from temporal distance serves a limited purpose as a hypothesis because the effects of conscious perception as an ability in matter is not well understood and the presence of consciousness in all matter is also mysterious. I do not, however, disregard such a possibility of a more latent connection between all matter and objects. I leave this train of thought open. I acknowledge that not working with such realities may limit the relativity of these ideas I am writing.

These thoughts aside, there are fundamentals to the tendencies of morphological formulations in Rupert Sheldrake's writing. There is a very useful sentiment in the principles of potentiality and the possibility of morphological elaboration in all matter, consistent with Newton's first law of Thermodynamics of transforming energy, which states, "Energy cannot be destroyed only transformed."

As Sheldrake writes in *Morphic Resonance: The Nature of Formative Causation*, "In Newtonian physics, all causation was seen in terms of energy, the principle of movement and change. All moving things have energy—the kinetic energy of moving bodies, thermal vibration and electromagnetic radiation—and this energy can cause other things to move. Static things may also have energy—potential energy— due their tendency to move; they are static only because they are restrained by forces that oppose this tendency."

The natural systems act in a certain momentum. A particular tendency is in all natural systems that are shared, whether by laws, habits, or embedded processes. Human conscious willpower also interacts with such momentum of processes. Consciousness, perceptual awareness, or intelligence is an extremely controversial phenomena for understanding human agency as the human domain of influence has many implications. Natural systems move and behave with independent accord or perhaps more directly related to our behaviors and actions than we can possibly understand. Basic quantum experiments show that a particle will behave differently if we simply observe it, as opposed to when it's unobserved. Aside from these thoughts regarding the mind, consciousness, perception, and action from a distance, morphology as a method to understand and interact with nature remains a primary focus. Morphology then may be understood in the awareness of ecological inter-functionality and interdependency, that the morphology exists within.

A morphogenetic field surrounds all objects in our universe, in all scales, and levels of complexity, as a set of possible futures defined by the properties of that object and the properties of its surroundings. A morphological system independent from ecology can only be conceived in an ideal state as in a mathematical equation or an experimental laboratory, the possible reentry of the morphological system into an ecology can also be enacted—and interactions can be viewed in a detailed sophisticated way. Morphological systems reveal their properties to each other as in chemical reactions. A morphogenetic field is attributed to a particular object and its possible developmental elaborations as action system potentials become defined.

Sheldrake explains, "These fields order the systems with which they are associated by affecting events that, from an energetic point of view, appear to be indeterminate or probabilistic; they impose patterned restrictions on the energetic possible outcomes of physical processes."

The physical processes of morphology can be measured and understood as physical quality metrics, such as weight, volume, and rate of oxygen consumption.

"The progressive improvement of techniques permits such descriptions to be made in ever greater detail; for example, with the electron microscope the processes of cellular differentiation can be studied a far higher resolution than with the light microscope, enabling many new structures to be seen; the sensitive analytical methods of modern molecular biology enable changes in concentrations of specific molecules, including proteins and nucleic acids, to be measured in very small samples of tissues; by means of radioactive isotopes or fluorescent antibodies, chemical structures can be 'labeled' and 'traced'," writes Sheldrake.

The analytic experiments describe natural phenomena as morphological processes and classify behaviors to be compared with others in search for differentiation and resemblance in developmental stages, phases, trajectories, and tendencies. Different isolations and perturbations as parameters and variables will reveal consistencies and alternations. With many arcs of development recorded, identifiable causal connections for development can be proven. Sheldrake points out that parts of systems can be removed and their development observed and studied in isolation. He uses the example of grafts and transplantation. Transplantations of different parts will discern functions in anatomical relationships or capacities of interactions with different components and materials.

The morphogenetic process that is present both in the organic biological formulation and the crystallization materializations are conjured by energy being consolidated in the composition of dynamic forces. The dynamic forces that inform morphologies of natural systems have direction towards an ultimate state. Embedded in morphogenesis, as seen in the natural system, are the regulatory regenerative and reproductive processes that always perform in methods that respond to the functioning of the whole system. *Entelechy,* a Greek word, meaning "the realization of potential" is the principle that guides the development and functioning of an organization or system. Its derivation (en-telos) indicates something that bears its end or goal in itself; it contains the goal toward a system under its control.

The built-environment then, instead of being an unpredictable amassing of sprawls, should meet a coordinated sophistication of the natural system as its totality is continuously referenced in its morphological formulation as a terrestrial architecture.

Our built-environment architectures are quite clearly composed of substrates for conduits of energy to move through. The physiognomy of the Earth is present with pathways of eroding rivers and estuaries that create fertile life. An abundant array of many climates for different forms of life all with different energy compositions relative to the morphology of the landscape and the biologies are present within the environment. Forests, rain forests and grassy plains distribute energy in alternative forms of morphology. Cities in infrastructural pathways for transportation and hydrology move resources as dynamic energy fields, storing and moving energy along certain trajectories and vectors. One could at this point directly engage the entire conversation of morphology in artificial, ideal geometric and biological terms. Yet, still present, is the conversation of becoming familiar with the natural system through a process of comprehension. Only then, once the terms of interaction are established, can we elaborate in pure morphology. Ecology as an architecture of interconnected and interdependent systems and organisms can be expressed through mathematical formulas and formal systems that define the interdependent variables. The mathematical formulas would also have a geometric morphological expression. With computation, we can generate mathematical formulas as models of interdependence in the life web of the biosphere.

In the following chapters the process of comprehending the natural systems is continued through technologies that give us in-depth understanding of the natural world. The differentiation between the comprehension of nature and the exploration of morphology is important because not all constructs are natural and adhere to the energy dynamics of the natural systems. Certain morphological constructs may actually interrupt and limit the expression of natural energy. Fundamental morphological fields influence the composition of all matter and these present the opportunity to participate in an abundant universe that is ever expanding; it is truly full of infinite energy if interacted with properly.

Fields of energy are possibility spaces. As Sheldrake writes, "The possible ways in which the atoms can combine together are given the Schrodinger equation of quantum mechanics, which enables the orbitals of electrons to be calculated in terms of probabilities; in the quantum field theory of matter these orbitals can be regarded as structures within the electron positron field." The structures of the morphogenic field define developments of the morphology over time, providing order, sequence, limitations, and guides to the morphological formation.

The future potential of structures as a potentiality is held within a virtual future that is articulated by contextual information and internal qualities of the morphological, morphogenetic system. The virtuality of the morphology as a potential formational system is its possibility space. The actualization of a virtual potential is fundamental to the conception of a Terrestrial Architecture when environments must be conceived in a layered interdependency of interacting morphologies. Architect Greg Lynn in his book *Animate Form* discusses the concept of a virtuality becoming an actuality. He writes,

> "The term virtual has recently been so debased that it often simply refers to the digital space of computer-aided design. It is often used interchangeably with the term simulation. Simulation, unlike virtuality, is not intended as a diagram for a future possible concrete assemblage but instead a visual substitute. 'Virtual Reality' might describe architectural design but as it is used to describe a simulated environment it would better be replaced by "simulated reality" or "substitute reality." Thus, use of the term virtual here refers to an abstract scheme that has the possibility of becoming actualized, often in a variety of possible configurations. Since architects produce drawings of buildings and not buildings themselves, architecture, more than any other discipline, is involved with the production of virtual descriptions."

A multi-valence of any morphology is to interact with others. "Any given type of atom or molecule can take part in many different types of chemical reactions, and it is therefore the potential germ of many different morphogenetic fields. These fields could be thought of as possibilities "hovering" around it," writes Sheldrake.

Probability structures of morphological systems can exist at many scales and complexities. The temporal formulation of a natural system with phases and stages are opportunities for alteration and variation within familiar morphologies as the surrounding environment provides possible interactive derivations for alternative formations. For example, morphogenetics might be associated with the embedded code of the morphology as in its genes while morphogenesis may be understood as the initial proliferation of the morphological system. Morphology might be considered as the form, shape, or geometry of a phenomena in the universe; while the physiognomy and physiological structure of our universe assumes shapes and forms to perform certain functions which Sheldrake would call behaviors. The physiognomy of our environment is of prime importance for designers of the environment; to derive the nature of its morphology and to generate new methods of driving its formation is a duty and responsibility, as architects, developers, engineers, and artists are all generators of the environment.

CHAPTER 5

Fractals In Architecture

Many cultural value systems are part of global architecture. Industrial modernity was able to culturally appropriate itself in most major cities around the planet, in terms of steel construction and gridded skyscrapers for apartments and office buildings. Le Corbusier, Mies Van Der Rohe, Gropius, and other early modern architects adopted a method of designing and constructing rooted in principles of fabrication and mass-production. Prefabricated modular system components were able to be produced in industrial facilities and shipped to construction sites, and buildings would assemble themselves according to an industrial capitalist paradigm or the demands of the market.

The cultural, economic, and political models of cities demanded an industrial way to build many skyscrapers, warehouses, homes, etc., and the orthogonal and modular methods of construction that the early modern architects developed served the popular demand to create dense urban spaces with many market opportunities as businesses and services could be erected swiftly in proximity to each other.

As a cultural value system, industrial modernity has mutated and evolved into tech-industries, and we see hybrid companies like Tesla leveraging tech and manufacturing to create an ultra-capitalist archetype of fabrication and distributed design products. Mass production is alive and well in Nike shoe design, Galaxy phones, iPhones, and generic skyscrapers. Industrial design and product design might seem unrelated to architecture, but a culture of manufacturing applies to all manufacturing systems simultaneously. When aerospace technology developed software to build 747 airplanes, it was soon adopted for Frank Gehry's undulating surface spaces.

BIM and digital modeling now are at the forefront of delivering architectural projects. To illustrate the concept of culture and the contemporary manufacturing narrative of tech, consider the way an iPhone makes its journey to undeveloped areas; whereas, sophisticated architecture does not, so the narrative of tech manufacturing is perhaps more relatable than contemporary building practice narratives.

The value systems that drive design and manufacturing in the tech industry function in terms like utility, efficiency, and innovation, so-called "optimizing" our human condition. The motive behind much of recent production is predicated upon user adoption. Beyond this scope of intention there has not been any additional criteria to define our design/manufacturing profession. The contemporary design manufacturing industry is predicated upon the individual needs of users who can afford the products. Most architectural projects follow a similar paradigm, in a similar way, creating luxurious experiences or elite functional spaces available to be leveraged by the rich. Contemporary architects are client driven; clients are driving what architects are building and clients' motives are often market driven or economically driven. The correct design pedagogy and practice paradigm for our current time on Earth, that responds to the major design questions we are facing on this planet, is likely not going to be defined by the highest bidder alone, who is seeking to create the most hype or "profitable" model. However, the correct design pedagogy and practice will have a value system and will likely be profitable. This can be defined by collaborative teams of interdisciplinary intelligences that can formulate new modalities of environments.

Fractal Architecture is responding directly to a different model of societal needs, which is more holistic as it addresses the environment and architecture in a greater, more comprehensive, inclusive totality than the immediate needs of select individuals. Architecture can no longer serve the motives of finance and public and private power alone. Greater principles were in place far before our socio-political power structure scenarios.

Mathematics is the mother of all sciences. It outlines geometric systems that apply to every form of science, including physics, chemistry, biology, etc. Magnetic fields are represented by vector field mathematics. Biological cellular propagation and subdivision is best understood through principles of exponential growth and radial crystal lattices. Crystallography defines the growth of crystals through axes of symmetry within a sphere to identify cubic, quadratic, isometric, tetragonal systems. The forefront of chemistry is being defined in the geometric architecture and structure of molecules to create new molecules by modifying their

NASA James Webb Telescope Photograph of Dancing Galaxies

structure, such as in the work of Omar Yaghi, as he harvests water from dry air. The list of mathematical formulations that define science is endless, and most often the formulas have geometric morphological phenomena that are visible through simulation and instrumentation.

The universe and biosphere precede humanity with built-in axioms or mathematical constructs that define the behavior of it all. We are immersed in a reality that is composed of atoms and molecules, and **59** atoms are 99.9% empty space. Atoms are composed mostly of the fields through which electrons travel.

Why is this important for architecture?

If we architects manipulate the material world to construct buildings, cities, worlds, urban scenarios, and landscapes, we must be fundamentally aware and conscious of this "material world" that we are manipulating, which may be understood as an electron negative field space which is more like a hologram as it is mostly hollow and empty. The definition of a hologram according to Google is "a three-dimensional image formed by the interference of light beams from a laser or other coherent light source."

Science, architecture, and conscious intelligence is ever increasing our ability to manipulate the material world hologram. Architecture and engineering has progressed from the primitive hut, to urban blocks, to city master plans, to continental infrastructures, with the rise of geospatial technology, which allows us to scan the Earth and make digital models of it. We are on the brink of defining the physical structure and chemical composition of the planet. As we become familiar with the reality we inhabit as a mathematical play of variables and forces, we recognize a consistent pattern across its inner workings.

Mathematician and author Douglas Hofstadter, in his book *Gödel, Escher, Bach: An Eternal Golden Braid,* outlines overarching mathematical principles that apply to every branch of intelligence imaginable. In the musical compositions of Bach, Hofstadter identifies algorithms, the formal mathematical systems of logic that are the foundation for computer coding and artificial intelligence. He traces the origin of geometry, finding it in the exact same place as the origin of mathematical proofs with Euclid, with Euclidean Geometry and Euclid's proof. Hofstadter defines infinity, types of infinity, how infinity is expressed geometrically and algorithmically and the isomorphisms of infinity-types, which correlates to how the mathematics of infinite geometric systems apply to all conditions of the universe that we find and experience, including the structure of our DNA. Principles of Recursive Fractal Structures are used to explain abstract mathematical realities that apply to all branches of science.

Without explaining every application of Hofstader's thinking and the all encompassing relevancy of recursive fractal structures, I can offer a few immediate pertinent fundamental applications of these principles for architecture and design. The multi-scalar nature of a fractal algorithm allows for a fractal system to propagate at every scale in a unified way. We can consider a mycelium network of fungi, linked to the root structures of trees in a forest. The biological cellular nature of both the fungi and the trees are an overarching cellular structure that allows for the transmutation of solar energy through photosynthesis and the absorption of nutrient minerals from the soil, and water also.

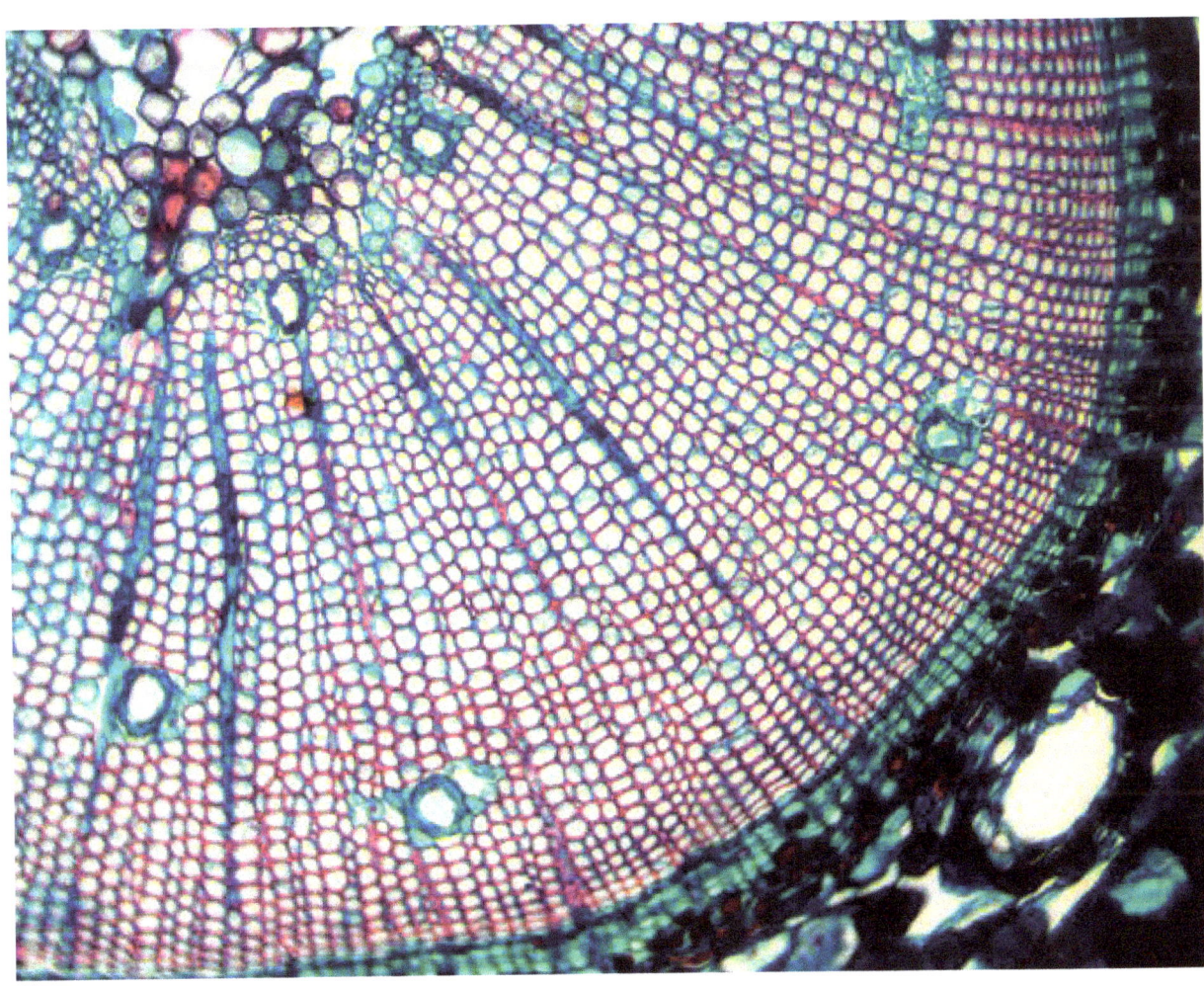

Cross Section Pine Stem Under Microscope: Creative Commons

A Photograph of a Forest on Orcas Island, Washington: Jack Oliva-Rendler

Comparing this multi-scalar cellular diffusion of a forest to a city would involve considering the city's overarching mobility grid that is a schematic for social programs and its relationship to the topography of the landscape. For example, an interior grid of a building is an instance of expression, one expression of a larger system. The interior grid of buildings is continuous with the grid of the city (just like trees are continuous in the grid of the forest). Together, they form a matrix of subdivisions of space. In a city, the parceled land and partitioned buildings form a modernist grid fractal lattice like a matrix, with addresses and attributes for each component. This is an important point because the unique buildings and the city grid function in a different sense when viewed as a continuous fractal entity. This might be referred to as the urban fabric—continuously subdivided cellular systems that are driven by mathematical formulae that create fractal matrices. It is a technical description of a fractal city and each building is an individual expression of the overall formula of the city.

Considering that there may be an algorithmic formula to explain the structures of the forest and its cellular diffusion as biochemical morphology, we should then also be able to construct formal system algorithms to define the multi-scalar composition of building components in our built-environments. This is purely from a schematic functional perspective without considering the psychological, emotional, spiritual, and conscious states that would be triggered by existing within a harmoniously-structured environment. I have proven that at least schematically and conceptually in a barebones proof of concept sense that a megastructure can be constructed by a mathematical formula that retains plenty of diversity in architectural elements in terms of bridges, columns, floor slabs, volumes and hallways rooms, entire buildings, highways, chambers, structures within structures, etc. The following images are renderings I have made from within a singular fractal structure generated by one formula in Mandelbulb 3D software that creates a seemingly infinite variety of spatial conditions that are linked in a potential infinite city structure.

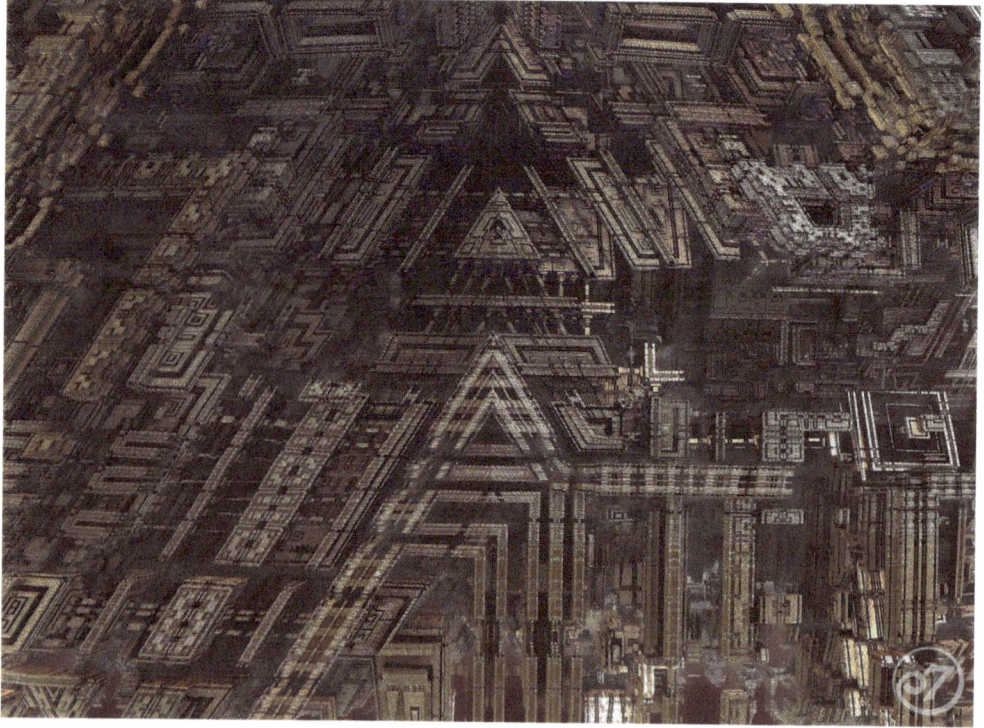

This megastructural city is modular or cellular. Each piece and element is connected to others in adjacency and alignment. Large complex structures are often composed of simple modules that mutate and adapt as they are diffused and propagated across a system; a recursive structure always retains a root rule system, even if the rule system is evolving.

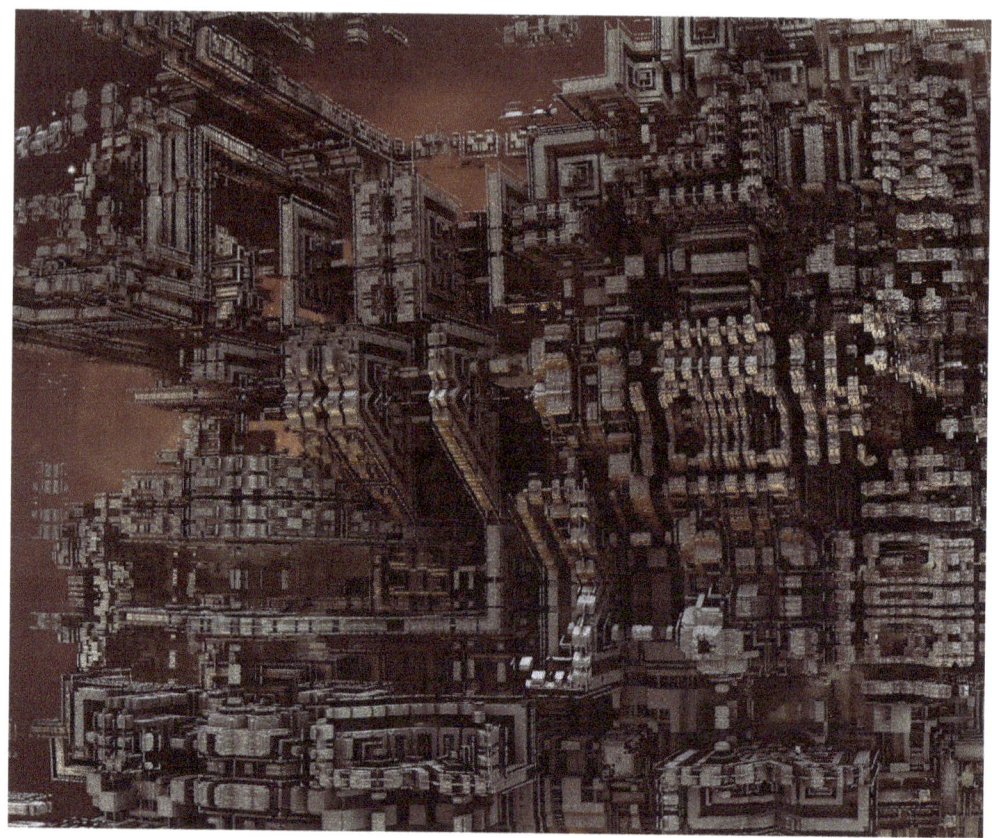

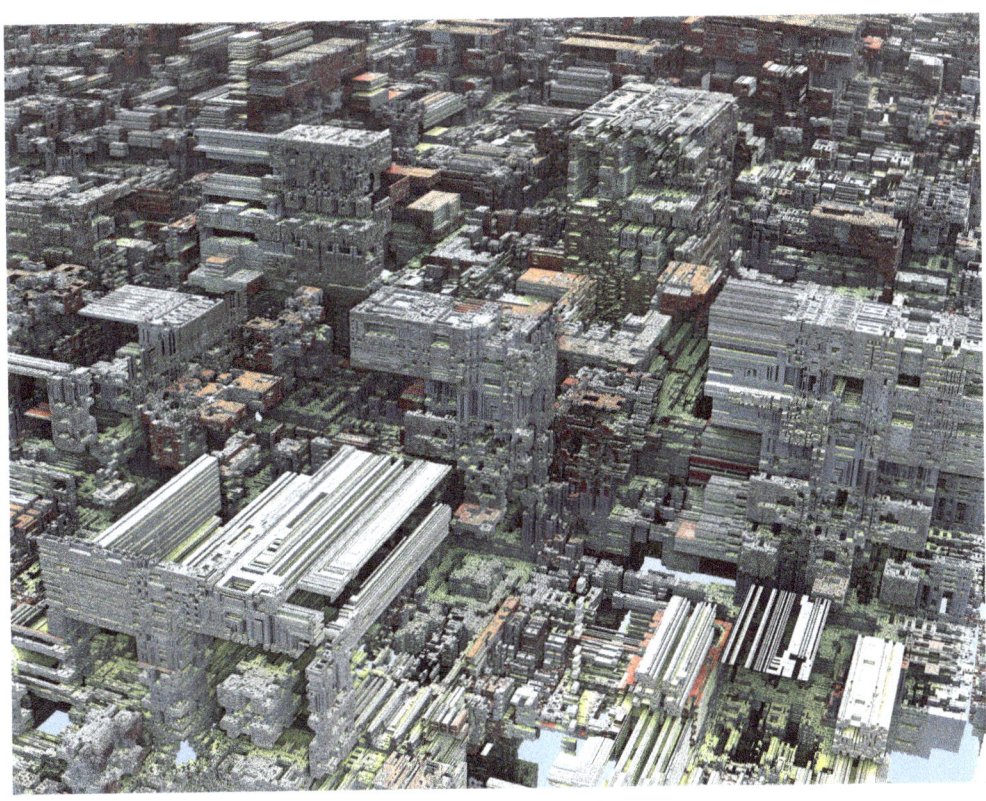

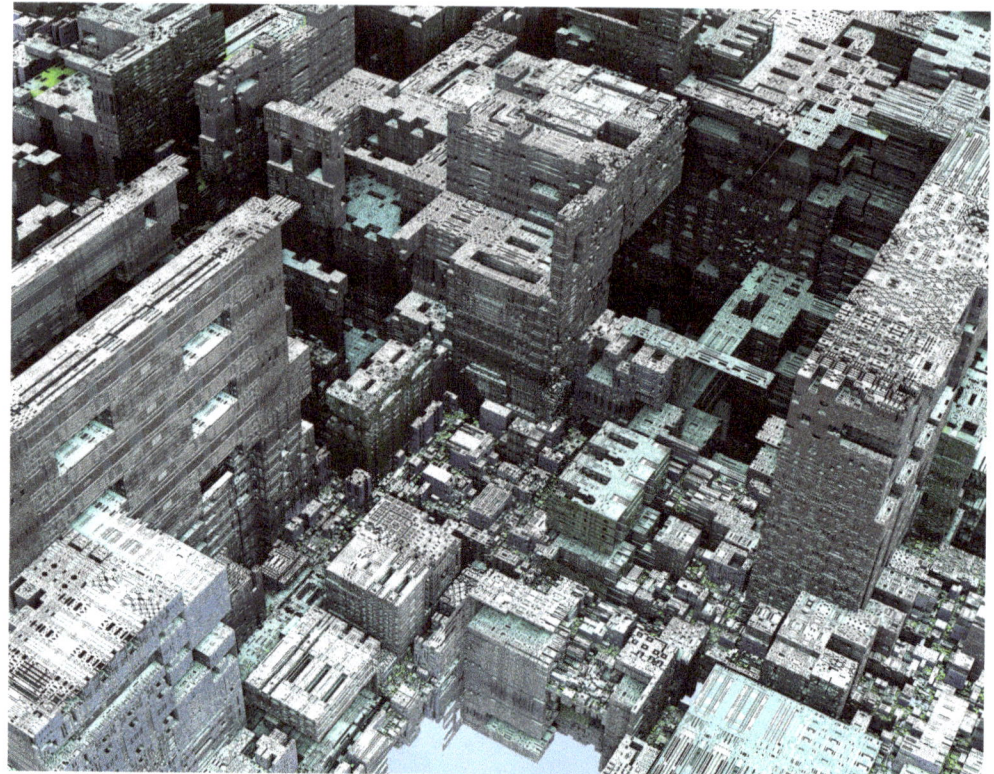

This individual structure, that I constructed with my fellow Futurly instructor colleague Mumin from Austria, is a process that can be scaled to infinity also, as surfaces follow recursive surface tectonic mutations. The individual structures are elemental components to a larger infinite system, like a cell within a living organism ready to multiply, diffuse, and propagate.

Fractal algorithmic mathematical forms allow for an ability to assemble complex structures with a degree of organization that brings

coherency to a composition. Without an underlying rule set or algorithm, complex structures will quickly become chaotic and disorganized. Rule systems can be sophisticated and elaborate providing great degrees of freedom and diversity while still offering structural compositional integrity.

Fractal algorithmic geometries are foundational for the future of architecture as we transition into a fully digital workflow to manage our built-environments. Architecture has always involved calculations to achieve desired results and computation is an extension of mathematical calculation. Because eight billion inhabitants live on Earth, we will absolutely require tools to plan our harmonic coexistence. Algorithmic design principles will incorporate the cooperative functioning components of our environments into well distributed structures that meet the needs of the ecosystem. Schematic fractal geometric systems can organize how multiple infrastructural and programmatic functions coexist within the same scheme offering all the societal and ecological needs within a multi-layered interdependent synergetic algorithmic structure.

The technicality of constructing fractals isn't rooted in one particular software; although, there are some dominant computational forces at play in the contemporary tool kit. Fractal construction is more a heuristic thought process than a technical procedure. A few prominent abilities come from a sensibility for formal systems and algorithmic processes. Complex sets of data (or information that are pertinent for driving design decisions) are no longer overwhelming. We are able to organize referential information into relative structures that translate spatially. The associative nature of adjacent networks and branching generations of related space is conducive to understanding how we can organize the complex demands of our environments. We can assess what the criteria are for a design, the fundamental parameters and considerations, and how the parameters as a base-code index drives the design decision making, as the parameters are transmogrified onto formal systems, and schematic fractal geometries.

Fractals are four or more dimensions. They propagate, diffuse, expand, mutate, and adapt, transforming as their base-code and variables evolve. As a diffusion and propagation system, the fractal schematic forms allow for underlying organization as programs and functional elements are assimilated into a schematic plan. Orthogonal grids are often used for modernist schematic plans which are already modular and cellular as a fractal would be. As an abstract geometrical schematic, fractal recursive structures would be more of an evolution to the modernist grid than a radical antithesis. In the fractal recursive structure scheme of geometry, there would be a procedural way to subdivide and multiply cells within a grid, also adding variations of angles and articulations. One can compare Eisenman's classic diagram of a nine square cube with a tessellating fractal cube. Eisenman's designs for his house are an index of geometrical procedures, a multiplication and division process of space. Although Eisenman was positing the novel intellectuality of the cube, I would add that the geometric capacities are capable of consolidating and organizing space to achieve programmatic and engineering functions also.

A generic fractal image may resemble the plan for Versailles and also Barcelona. All these geometric systems are axial and cellular with multiplication of lines and subdivision of space with proportions and alignments remaining consistent.

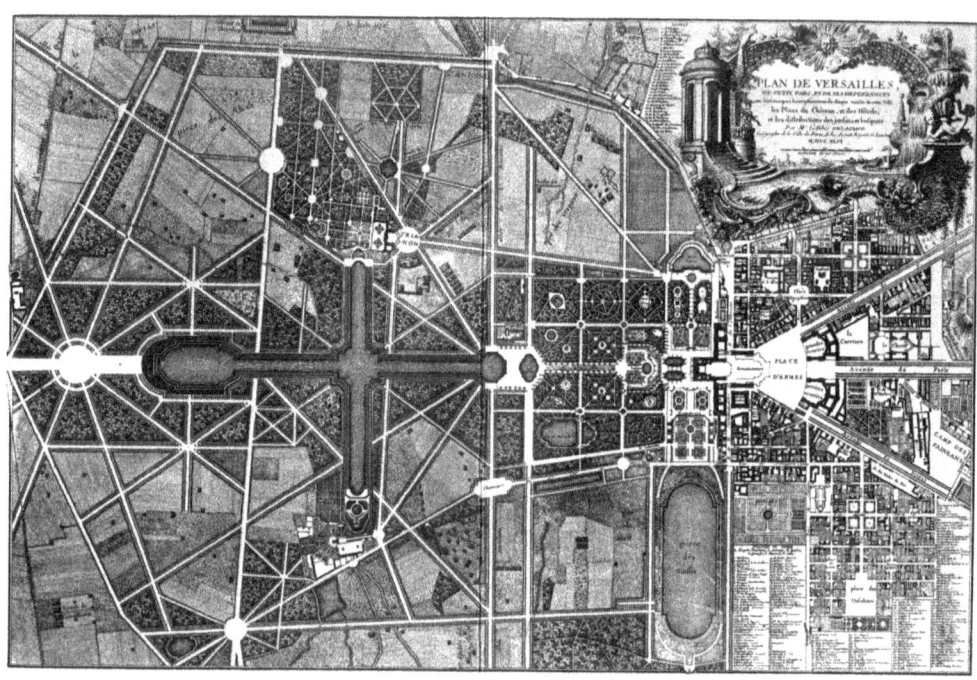

Versailles and gardens from 17th/18th century

At the scale of the house the sub divisional offsets of space create adjacencies, proximetes, and major axes to traverse the program. Volumes are continuously fragmented or fractured as in a volume forming incomplete enclosures that are nested within each other. By outlining the major volumes and highlighting the way they nest within each other, we discover an abstract geometrical system based on the subdivision of rectangles. Subdivision is a subset of the concept of division, which is the basic unit of a square root and correspondent and opposite to multiplication, which is the basic unit of an exponential formula; all such mathematical operations are basic fundamentals to fractal structures.

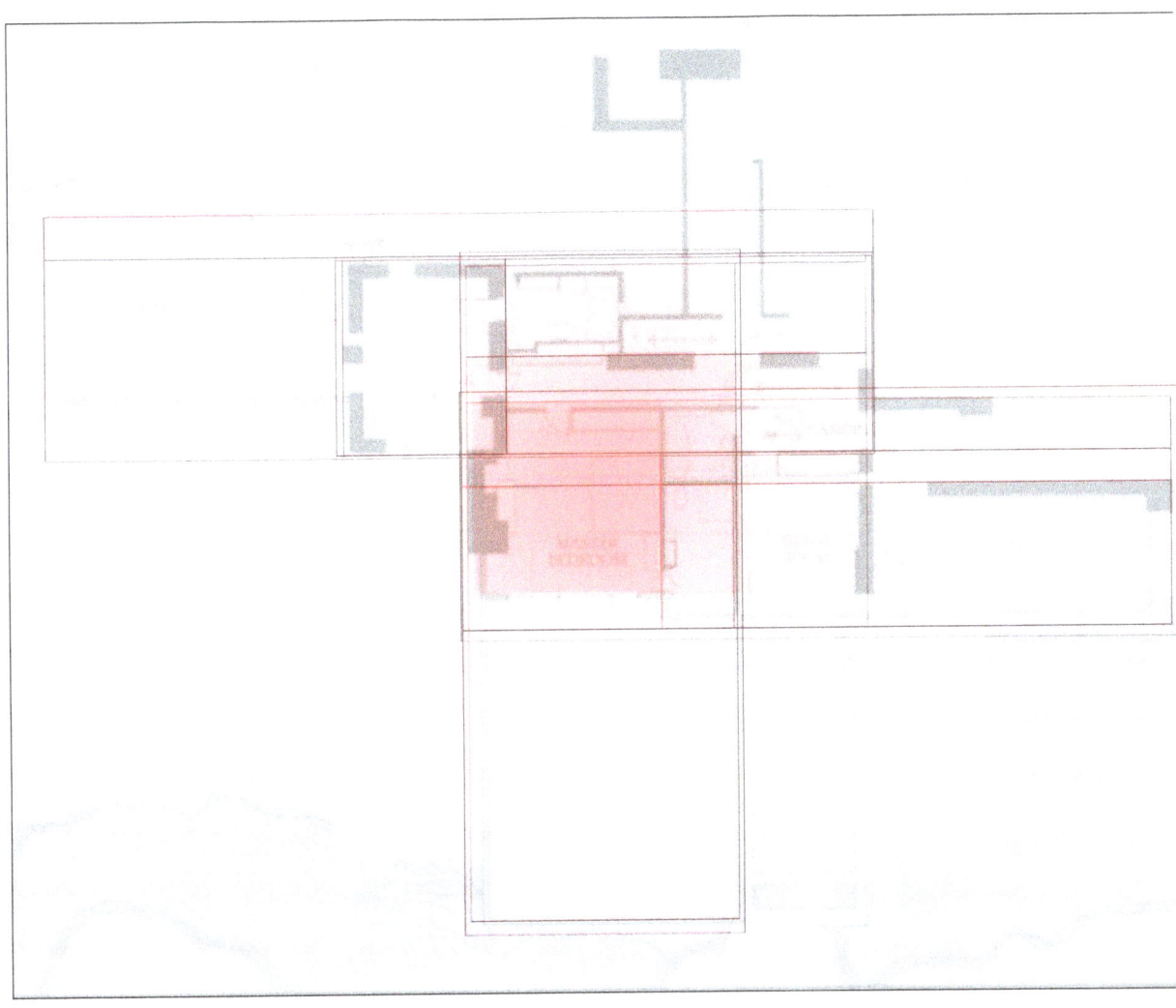

Second Floor Plan - Fallingwater, State Route 381 Spatial Study - Diagram that I constructed

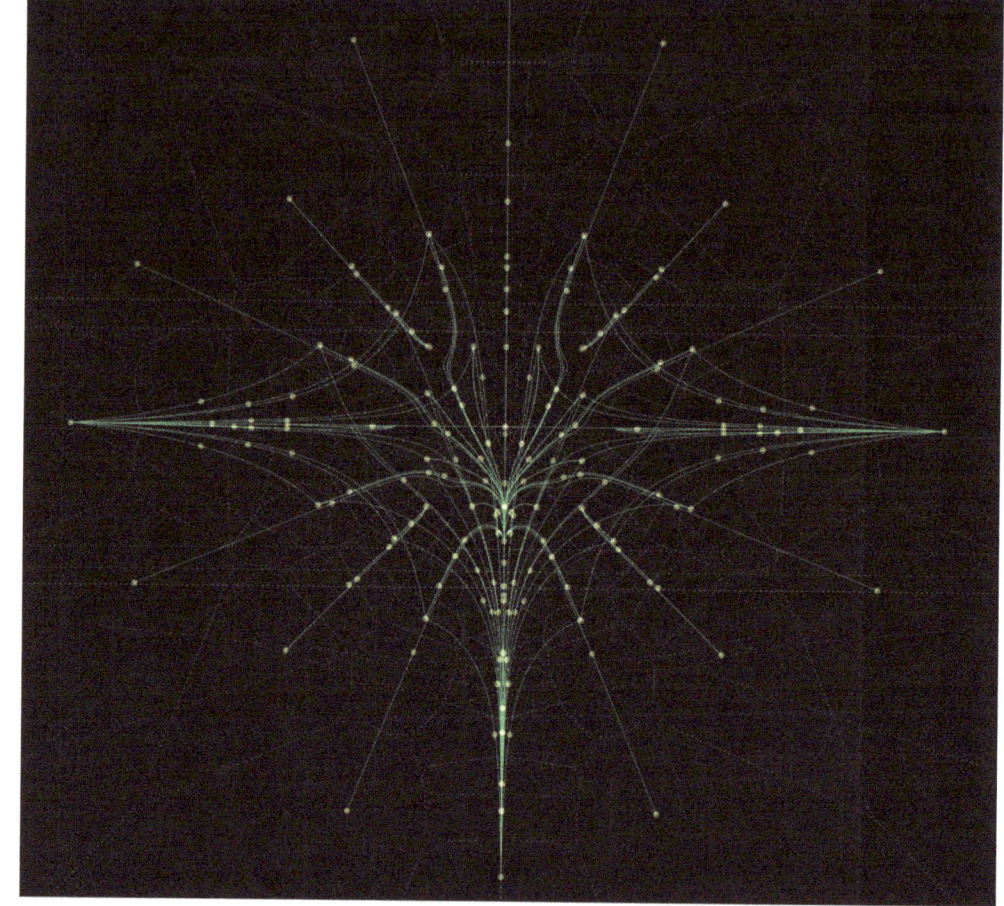

Fractal field algorithms can allocate locations and flow lines for ecological systems, such as irrigation and hydrology paths and seed casting distributive patterns. The same parametric understanding of fractal systems allows for the construction of landscape models, here sharing a model of the north american continent generated by an algorithm that subdivides topographic contours and generates a mesh model. This 3D topography model was constructed during my time as a design consultant for Metabolic Studio in 2023 in Los Angeles.

USGS Digital Elevation Model

PART 2

Computation Relative to Terrestrial Architecture

The second part of this book explores technologies and their contextual implementations to the biospheric awareness written about in Part 1, the first five chapters of this book. Computation and cybernetics are explored to give greater sensorial awareness and insight into the natural systems that exist on Earth, including the ways we observe and transform the planet. The cyber-infrastructure provides a continuous relay of information generating geospatial databases that are constantly updated. The following chapters deconstruct, decode, and piece apart the cyber-infrastructure into layers, components, and philosophies in order to bring an in-depth understanding into computation, enabling new forms of design and participation in the Earth evolution. In a description of computation we begin to understand principles of calculation, as variables, formulas, and equations that represent dynamics of Earth realities, through displays of data, infographics, and multimedia types that are continuously evolving. A robust accumulation of data and information regarding ecology becomes the contextual information for project initiatives. The creation of the data and the organization data itself is a craftsmanship of its own; the organization of data is also a new contemporary form of curation and library architecture. These tools allow us to engage the complexity of the biosphere with precision at scale.

Part 2 of this book explores the future of our toolsets by analyzing our contemporary computational toolkit. As we articulate the potentials of our tools to spatially map and model our environments, new processes and potentials for design and visualization unfold before our eyes. The ability to engage multiple community projects simultaneously and cross-reference their interdependency is achieved. We are able to coordinate multiple models and projects within a large scope of land and initiate integrated solutions as the multiple layers and components come together in platforms and softwares. In Part 2 of *Terrestrial Architecture* design techniques are explained in terms of how they are able to address environmental and ecological complexity. Design techniques are translated into pedagogies, curriculum, and institutional frameworks to create opportunities for new ways of designing and thinking in regards to scalar and dimensional shifts. Large infrastructures and natural systems begin to merge into agents of terrestrial realities—we are able to view the intricacies and functionalities of large scale terrestrial systems and participate with agency in their evolution.

At both the Southern California Institute of Architecture and Harvard University Graduate School of Design, I presented thesis projects, "A.T.L.A.S: Apparatus for Technical Logic in Analytic Structures" and "Instruments of Terrestrial Transformation," about large scale buildings that served as data infrastructure environmental components. These two projects were related in many ways, twin to each other, reflecting and building upon each other in an exploration into accomplishing planetary-scaled architecture. The ability to write this book is credited to the back-end constructive components of the thesis projects. The book in its entirety is an extended and complete version of the thesis.

As an institutional framework, the thesis projects were structures that store environmental and ecological data, composing them into forms and media that are consumable and workable. As we are able to consolidate, organize, and visualize our Earth system information, we are then able to participate in profoundly expanded ways. Computation provides bits and forms of information; cartographies contain polygon lines with attributes and statistics embedded in them. The geometry of spatial constructs are documented with coordinated precision and multi-scalar contextual relevance. The multitude of maps, models, simulations, and multimedia components that are necessary for a Terrestrial Architecture requires an architecture to organize the information and memory systems. As we build our data infrastructures and engage the complex reality of the biosphere—designing and intervening on the behalf of biodiversity and the beings we care for—it may be of the utmost importance that we comprehend our technologies and how they proliferate our abilities. The samurai sees the sword as an

extension of his arm; the architect, following the technologies of remote sensing, digital modeling, and geospatial cartography, sees his toolkit as an ever-expanding capacity to touch the Earth in a rippling gesture of terraformation.

CHAPTER 6

Cybernetics

Cybernetic computation happens through mathematical formulae, but their theoretical and conceptual performances can be described as a process of memory, information storage, and manipulation in alternative descriptions. Computers store and manipulate information, derive patterns, calculate relationships, and ultimately interpolate futures of causal mechanisms behind complex systems. The fundamental concept of computation can be regarded in the act of interpolation, understood as the deriving of a curve, implied by a set of points in space. The points serve as an informational basis to describe the curve and the points along the curve in space. This was useful during the development of cybernetics since its aim was to create a new type of warfare, requiring the calculation of the future position of air \planes as a means to shoot them out of the sky. The airplane would form a curve in space as it moves and the computer would calculate its trajectory, its movement. Norbert Weiner refers to this in his book *Cybernetics: Or Control and Communication in the Animal and the Machine* when he writes, "To predict the future of a curve is to carry out a certain operation on its past."

The calculation of positions evolves into greater informational constructs that discern the ability of prediction and awareness of adjacent positions in time and space or the virtual potential of a force. Communicative sensory awareness between surroundings and the organism or the machine are feedback loops. Feedback loops are essential to be deliberate about decisions that have a causal relationship with our surroundings. For example, our nervous system, from our eyes, to our fingertips, to our visual cortex inside our brain constantly relay information of what is being sensed, describing and articulating the reality in front of us, enabling a coordination of our movements and thoughts. Similarly, the sensory systems of cybernetics function through signals as messages that occur in a time-series, as statisticians call them. There is a certain periodic scanning of information to retain an existing model of what is encountered called the time sweep, and in the human body and its

85

feedback system it is known as the "alpha rhythm" of the brain. Cybernetics understands the ultra-rapid computing machine principally as an ideal central nervous system.

What is being recorded and measured by the cybernetic organisms and machines is a passing of time, a phenomenon that occurs in relation to the passing of time. Time could be measured by the movement of planets in relevance to each other in a definite exactitude as the movements follow laws of motion and mechanics that can be reversed.

Weiner writes, "Thus if we were to take a motion picture of the planets, speeded up to show a perceptible picture of activity, and were to run the film backward, it would still be a possible picture of planets conforming to Newtonian mechanics. On the other hand, if we were to take a motion-picture photograph of the turbulence of the clouds in a thunderhead and reverse it, it would look altogether wrong. We should see downdrafts where we expect updrafts, turbulence growing coarse in texture, lightning preceding instead of following the changes of cloud which usually precede it, and so indefinitely. The term 'cloud,' 'temperature,' 'turbulence,' etc. are all terms referring not to one single physical situation but to a distribution of possible situations of which only one actual case is realized."

Cybernetic communication and control machines define a new awareness of informational dynamics within our universe and give us analytical awareness of the processes of our ecology, even in real time. Our ability to simulate and model defines a new paradigm for design and a new scope and fidelity as a reference point, that is the scale of the environment and the processes of the biosphere and social ecology. It is an ability to monitor, simulate, control, and understand groups of transformations simultaneously, so we can derive a virtual space consisting of possible futures of our living systems. Such an ability to sense environmental phenomena is articulated by an infrastructure of sensory instruments, where if there are muted signals and messages or a lack of ability to organize and integrate sensory data the result will be discoordinated movements or actions, as a body with a damaged nervous system. The following is an example of a feedback mechanism: a ship being controlled by a gyrocompass, referential to a current direction and a set course from the quartermaster.

Weiner writes, "In the important book of MacColl, we have an example of a complicated system which can be stabilized by two feedbacks but not by one. It concerns the steering of a ship by a gyrocompass. The angle between the course set by the quartermaster and that shown by the compass expresses itself in the turning of the rudder which, in view of the headway of the ship produces a turning moment which serves to change the course of the ship in such a way that the turning velocity of the rudder is proportional to the deviation of the ship from this course, let us note that the angular position of the rudder is roughly proportional to the turning moment of the ship and thus to its angular acceleration."

Informative feedback as ability to anticipate the current or future position and state of a system is a procedure that can be put into practice. Homeostasis, as mentioned in the description of the biosphere like we talked about in Chapter Two, is a feedback process.

Infrastructural homeostasis is a series of relays that occur in a periodic scanning discerning existing chemical or calculable states of information that are clocked at certain positions or states within the transformation, which may trigger reactions to balance or stabilize a system. Each neuron receives and communicates with a network of other neurons and their contact is known as synapses. "A very important function of the nervous system, and, as we have said, a function equally in demand for computing machines, is that of memory, the ability to preserve the results of past operations for use in the future," writes Weiner. Memory storage of different kinds serve different purposes, either for circumstantial current processes or memory usage that occurs in every action as a permanent record to be integrated. The content of the stored memory is influential on how it is to be used—contextualizing the memory within a network of other memories. Forms of memory recall are discussed in the field of cybernetics, such as group scanning and periodic scanning of memory systems.

Described in the architecture of a memory system to perform a group scanning is an array of ordered sensory leads that are coordinated

87

in their alignment. Spatialized is the reception of stimuli and the ability to store the memory of perception is embedded not in one neuron but in the constellation of receptors that have articulated relationships with each other. The relationship of receptors in their coordinated firing and receiving enables the pattern recognition or the grouping of relevant information as the information is compatible and important among a given set of receptors. Here, the nervous system is used as an analogy for an infrastructure of remote sensing devices and the processing of their data. As we discuss synaptic architecture of feedback relays and sweeping scans of energy exchange, we begin to articulate a process by which we are able to comprehend, visualize, and interact with our surrounding world.

CHAPTER 7

Cyberinfrastructure and NASA's Earth Observing System

Generating simulation models or representations of the Earth and its systems can be oriented around the satellite infrastructure as a remote sensing apparatus that creates a large database of information regarding the Earth. The remote sensing infrastructure consists of landbased laboratories and vehicles, drones, airplanes and satellites, virtually any device that can transmit sensory data remotely and autonomously. Different generations of satellite infrastructure projects have achieved levels of fidelity and their sensory information has been categorized for defining the legibility of certain terrestrial systems, including agricultural systems or urban areas. Through variant technologies that have evolved, the earth has been photographed and scanned from outer space using Geometric Orientation Technology, Photogrammetry and Spatio-Triangulation, etc. Photogrammetry uses Ground Control Points (GCPs) to create more accurate models or image data orientations. In *The Photogrammetric Record* article titled "Fully Automatic Generation of Geoinformation Products with Chinese ZY-3 Satellite Imagery," Zhang and colleagues write, "To achieve a more stable orientation result in sparse control areas such as seas and forests, spatial triangulation based on strip images is adopted (a single strip image typically covers about 50 km by 4000 km on the earth's surface)." Dense image mapping based on multiple images is used to extract object coordinates from dense points and feature lines. Then a dimensional elevation map (DEM) is generated. The DEM is created by orthorectification and color fusion based on the original image. The Dimensional Elevation Map serves as a topological and topographical model, representing the landscape in digital space. As we look at data types and information formats the applications of aerial photography and remote sensing via satellites and flying devices are vast and diverse.

Remote sensing instruments and aerial cameras, devices of terrestrial measurement and analytics, document information about the biosphere and the built environment, creating useful data for geology, soil mapping, agriculture, and forestry. Airphoto interpretation can identify resources for extraction, minerals, vegetation and can classify crops—timing the position of the plants in their life-cycle phases: preplanting, plowing and planting, emergence, mid growing season, preharvest, harvest, postharvest. In forestry, timber may be managed, harvests may be planned and coordinated with healthy growth rates to avoid mass consumption. Airphotos can monitor logging and reforestation, plan herbicide usage and the addition of fertilizer, and assess the health of forest nurseries, addressing fire potentials and suppressions. The images can detect slope failures and soil erosion. Wildlife may be censused and power line locations may be mapped and planned. Water retention strategies may be identified and the planning of irrigation patterns may be developed, enabling the use of water for power generation, drinking, manufacturing, or recreation. Urban uses include identifying land cover and land use suitability, lot and and house sizes, building setbacks, street widths and conditions, and curb and sidewalk conditions. Open space may be maintained and proximity to parkland or industrial land use may be measured. Transportation route locationing and infrastructural planning happen by the use of airphotos. Landform identification establishes multiple drainage patterns, soil textures, and susceptibility to flooding. The ability to identify such qualities is derived in the skill of the observer or machine intelligence that is trained to understand and use the photos for interpretation and identification.

An array of sensors may be attached to the remote sensing devices, multispectral, thermal and hyperspectral scanning, such as ground penetrating radars enabling cross track thermal scanning, thermal radiations and aquifer and underground infrastructure identification. Digital imaging post processing opens possibilities of uses, through image classification, data merging, and GIS integration, biophysical modeling, geometric correction, contrast manipulation, contrast stretching, spatial filtering, and edge enhancement. Different instruments can be programmed to observe and record specific terrestrial phenomena.

92   Here are some examples:

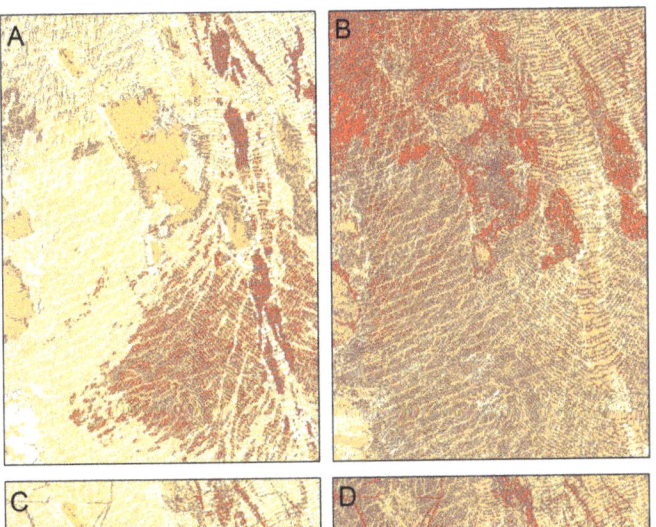

Above: USGC "Geiger Mode Lidar Over Chicago"

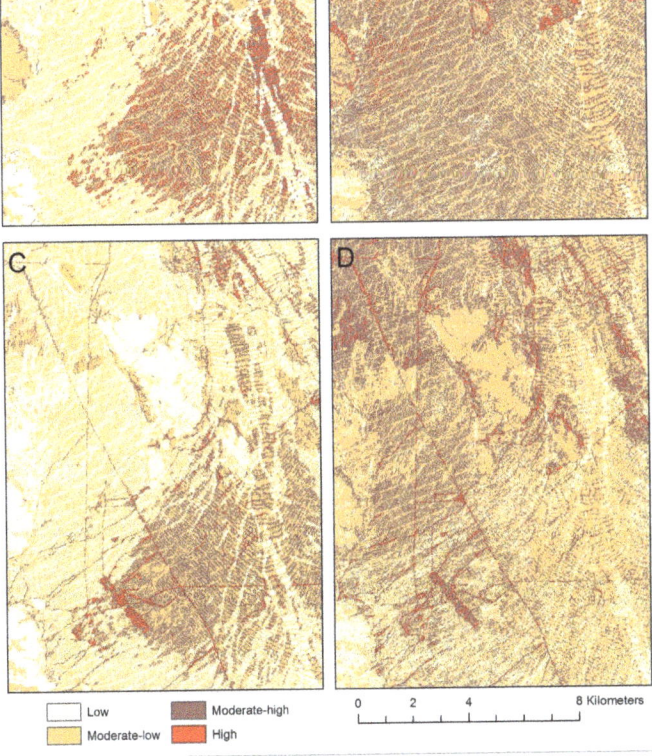

Low
Moderate-low
Moderate-high
High

0 2 4 8 Kilometers

Left: USGS Remote Sensing for Soil Erosion How Remote Sensing Was Used to Detect Soil Erosion

USGS Lansat

SAT Image Phoenix, Arizona

One example of a NASA-based program that observed the Earth is found in the book *Land Remote Sensing and Global Environmental Change: NASA's Earth Observing System and the Science of ASTER and MODIS.* The editors write: "The Earth Observing System (EOS) mission consists of a space-based Earth observing system, a data and information system (EOSDIS), and a scientific research program. Three successful EOS satellite platforms were launched between December 1999 and July 2004, and remain operational to date which provide the first coordinated and simultaneous measurements of the interactions of the oceans, atmosphere, solid Earth, and hydrological and biogeochemical cycles. NASA's EOSDIS is the fundamental infrastructure to support the EOS science mission."

The EOSDIS also provides the framework for current and future international cooperative efforts, like the Global Earth Observing System of Systems, or GEOSS. Such data infrastructure becomes available and used by interdisciplinary interests, yet postulated in this book is a focus on the built and natural environment. Educational, institutional, and business infrastructures that are involved in architecture and engineering projects will use virtual platforms to access the geospatial data and begin to seamlessly articulate the built environment with the processes of ecology and the biosphere as they are documented and described in the data infrastructure to create conditions for life. Platforms and institutional architectures to manage data accumulations are the pinnacle project of cyberinfrastructure. Through such platforms of data aggregations, either as networks accessed remotely or infrastructures embedded institutionally, architectures of the terrestrial scale will be formed and implemented with energy relationships that accumulate on geological, terrestrial, and atmospheric scales. Reliable and disciplined infrastructures for data processing are required to ensure long-term transformational results. The change will be measurable and drivable through the cybernetic infrastructure.

The origin of environmental data is generated by remote sensing devices that are launched into space or aerial photography. Aerial photography began with the V2 rockets and the aerial photographs captured from them. This was the beginning of a new vantage point to view Earth from above, illustrating a terrestrial reality from a new perspective. Aerial photography evolved into remote sensing devices from which we constructed a geospatial instrumentation network. This was a radical evolution for cartography as a map making craft.

First Image of Earth from Space:

Picture taken from a V-2 rocket over New Mexico in 1947

According to NASA's *The Earth Observer: Perspectives on EOS, as written about in Land Remote Sensing and Global Environmental Change: NASA's Observing System and the Science of ASTER and MODIS,* "The EOSDIS data centers became known as Distributed Active Archive Centers (DAACs), which became the implementers and conveyors of key practices and responsibilities at this time to ensure that data products were well documented, maintained in easily available form, and supplied to those who requested them. EOSDIS' vision soon started developing into reality."

Data accumulated in the satellite program and generated databases available to interdisciplinary usage, such as the U.S. Geological Survey. Open GIS and Open Geospatial Consortium are likely agents to define the future use of such data collections. Open GIS and its integration into educational, business, and institutional platforms as workflows that orient themselves around compositions of data inform a domain of architecture to be produced in relation to the environments that are being computed. Here are some more examples:

USGS: Chile   *USGS: Kalgoorlie Gold Mine, Australia*

USGS: *Yanacocha Mine, Peru*

USGS: *Chuquicamata Mine, Chile*

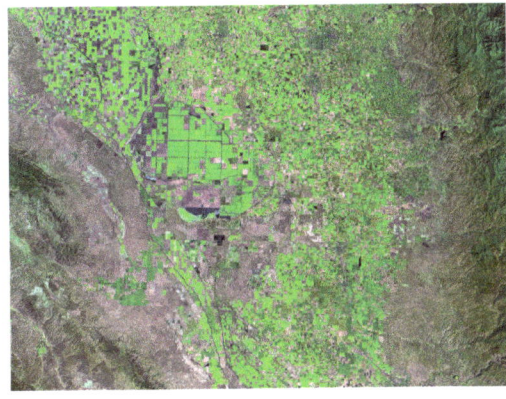

USGS: *San Joaquin Valley, California*

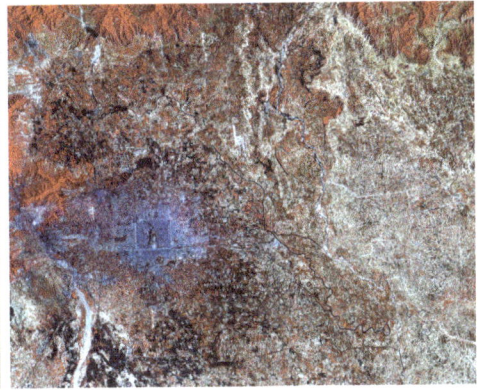

USGS: *Beijing's Growth*

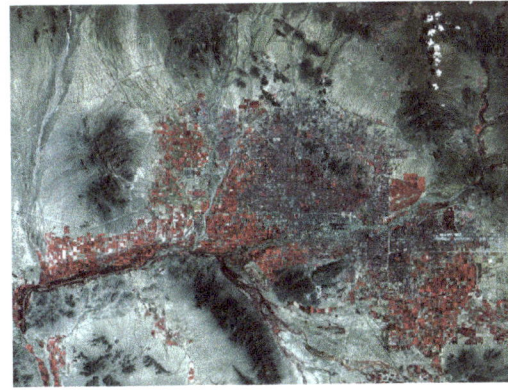

USGS: *Phoenix, Arizona*

USGS: *Sydney, Australia*

Cape Canaveral, Florida Image Source: Kennedy Space Center from Landsat 7

Triangulation and documentation of the landscape is a process derived in the history of cartography to dimensionalize, measure, and quantify the space of a territory or terrestrial plane. The geometric

98

translation and encoding of such information enables manipulation and augmentation and simultaneously provides the sensibility of familiarity to the plane's existing processes that are inherent and existing either politically or ecologically. Such proliferation of geometry across the terrestrial plane as abstract representations of dimensional measurements are pivotal in deriving architectures of the terrestrial scale as algorithmic design principles allow the ability to control and comprehend large complex adaptive systems as networks or aggregates. The procedural process of algorithmic proliferations of geometry can be compared to the patterns of music, a paradigm of parametric architecture, or a history of proportional matrices that govern architectural schema in the history of architecture, as in the construction geometries of churches, cathedrals, mosques, temples, and ancient cities like Angkor Wat and Teotihuacan with geometrical ordinances to integrate multiple components and systems in a synergistic alignment of use and purpose. The comprehension and management of catalog data sets, informational matrices, geospatial organizations happens through platform architectures that engage the complexity and organize proliferations of information into composed aggregates for consumption and re-manipulation.

Organizations may be developed in tandem with Geospatial Technologies that provide remote sensing information like lidar scans and aerial photographs. Aerial instruments are produced by private industries, Lockheed Martin being a progenitor of remote sensing instruments used in the military, such specialized production includes softwares called vehicle control stations (VCS) a user interface communicates the navigation of Unmanned Aircraft Vehicles (UAV) instruments and the information that is being collected. Navigating spatialized intelligence is informed by data in the form of maps, images, and videos. Lockheed Martin developed hydrogen fusion tools as a geospatial information system that provides real time 3D world representations to give immediate context to information. Simultaneous location and mapping (SLAM) constructs stitch together and geo-register images and videos to enable real time immediate and immersive 3D reconstruction. With 3D models the Hydro Fusion tools can create a rendering of a site, like a construction zone.

We can create real-time structure from motion, such as the ability to generate 3D point clouds and 3D imagery in real-time as a vehicle flies; do site mapping, including comparing architectural drawings to actual 3D imagery to gather daily progress updates; apply the programming and tools to agriculture, including being able to map fields or forests for the analysis of plant health and cataloging; automate the inspection of large, critical infrastructure; survey wide areas of terrain; and covertly generate maps of evolving situations in real-time.

The U.S. Geological Survey as a library of ecological data collecting information provides the possibility of remote sensing missions or information gathering operations to comprehend the ecological systems and bring further fidelity and breadth to a geospatial archive. With an abundance—an ever-growing abundance—of geospatial information, libraries, archives, and virtual platforms provide accessibility and means of exchange become ever more necessary to create a world in which the architectural discipline and communities can engage the terrestrial architecture. A consortium of many businesses and agencies are fit to engage geospatial information through an Open Geospatial Consortium. Digital geospatial information is encoded into digital platforms for information processing, including storage, transmission, analysis, visualization.

Through interfaces, communities and individuals belonging to different agencies access geospatial data and perform operative techniques. The interface accesses a geospatial world consisting of languages that open possibilities to manipulability, organization, and metrics; geo-languages describe conceptual worlds as epistemologies of the real world. Coordinate geometry creates spaces that open detailed feature operations. Geometrical operations proliferate through multiple softwares that are familiar with the architectural disciplines and can be coordinated with geospatial technologies that provide large scopes of data fidelity and spatial breadth. Through a series of translations of digital interfaces that rearticulate geospatial and geometrical information, the simulation of the ecology and the city takes form and terrestrial architectures are designed.

For instance, urban conditions measured by the cyberinfrastructure may be expanded upon, but a current example includes a heat island effect from an asphalt terrain that has blanketed cities. Planning for a total surface area of a tree canopy would reverse this effect, change the air quality, and perhaps even create wind tunnels that regulate thermal transfer. Models of a city are generated from the data infrastructures; platforms that manage and exchange such models increase the malleability and access to environmental and social degradation as dynamics to be controlled and reversed.

Conceptual worlds that articulate a particular relationship of data and systems within the biosphere and the built environment exist with residual geospatial worlds. The geospatial interface provides different viewports for different specializations that orient and frame data, specifying purpose, scope, policies, functional decomposition, semantic usage, mechanisms and distributed interactions. The different viewports become interoperable as data retains similar structures and this enables a flow of information and communication between agencies domains and viewports. "Catalog Services" help users find information within the computational environment; they assist in the organization and management of data creating means of access. In the earth imagery case archives are "multiple petabytes in size; ingesting a terabyte per day." The processing of image semantics with automatic detection of features and geographic concepts is necessary.

This is why cataloging is used. Cataloging provides ways to arrange resource descriptions, mechanisms for access and retrieval of resources, a standard way of describing those resources, a place for users to find the information they are looking for, and multiple ways of reviewing the resources from a variety of formats, sources, and locations.

CHAPTER 8

Platform Architectures and Virtual Community Action Planning

Residual to the vast amounts of Earth system data created by cyberinfrastructures are platforms that allow us to engage the data in technical procedures that are user friendly. The design of a platform can proliferate the potential of the data being used. Earth system data is not monolithic; rather, it is characterized by its diversity, comprising various layers and types. This diversity arises from the multifaceted nature of the Earth's environment, which encompasses geological, atmospheric, oceanographic, and biological components, among others. Therefore, it is necessary to create data platforms capable of handling this complexity effectively. With many layers and components to the environment that exist in different data types, there are many platforms that have been created to organize, distribute, and work with the data. A platform may be conceived of as an interface or a managerial device that allows for the composition and decomposition of data. Our ability to manage projects and address the complexity of our environments may be centrally dependent on the architecture of our platforms that compose and organize our sequences of operations and methods of communicating. In the platforms I speak of, subject matter is discretized and organized into bits of information, digestible units with attributes and qualities, and composed by the user. Geospatial Information System Models create maps and 3D models of environments with layers and embedded informational components. The future of Terrestrial Architecture is found in how our environmental models and systems are cross referenced, disseminated, and broken down into multiple scales and worked upon in sequences of operation. In the future, Terrestrial Architecture will likely witness advancements in the integration of data from diverse sources. This integration will enable researchers and practitioners to develop comprehensive and nuanced models of Earth systems, transcending traditional disciplinary

103

boundaries. The ability to cross-reference and disseminate data will facilitate collaborative efforts to address complex environmental challenges.

A great reflection on platform architectures is a book by Benjamin Bratton titled *The Stack*, as in the stack in a computer as an analogy to represent this correspondence between a virtual model space, its city, and the earth. Bratton writes, "The Stack model suggests both the means and ends of a specific kind of platform sovereignty. It demands that we understand the designability of geography in relation to the designability of computation and to see the state (and other sovereign institutions) in relation to both at once."

The use of such a platform empowers each individual to contribute and participate in the transformation of the megalopolis, therefore liberated from titanic industries of government and private institutions that manipulate the built environment in discriminatory ways, creating embedded forms of disparity through discriminatory zoning. The platforms enable alternative forms of economics, geopolitics, philosophies, and ecologies. The manipulability of the city through an integrated software platform is authorship over the rights of access to spaces, ones that are designed for healing and interconnected ecological performance.

Bratton explains, "Platforms pull things together into temporary higher order aggregations and, in principle, add value both to what is brought into the platform and to the platform itself." Platforms are recompositions of actors into an infrastructure of information manipulation and treatment. Forms of communication are established between actors that create fluid flows of information in work processes, such workflows become embedded as the ontological and epistemological method of viewing the environment through the lens of the platform architecture. Bratton writes, "A platform's systems are composed of interfaces, protocols, visualizable data and strategic renderings of geography, time, landscapes and object fields." Platforms contextualize items of importance as informational components within themselves creating scenarios and situations that provide fertile and potent relationships and communications to occur. Their design is embedded with implied connectivity between agents as users and information as domain exchanges expanding horizons of interdisciplinary exchange and widening the scope of comprehension with compositional sophistication. The ability of what is to be envisioned and designed is proliferated in the platform architecture as curriculum and schedules assemble elements into compatible bursts of new forms of knowledge and creation. Reality itself begins to transform from ambiguous indescribable informational sound into comprehensible points of clarification and discernment to be translated into a dialectic response that is appropriate and largely and resiliently culturally relevant. Such platforms contextualize themselves with the biosphere and resource extraction processes as methods of translation to become architecture. A spectrum of synthetic and biological processes expands across the domains of creativity contemporary planetary context for architecture rises into view from the inside of an informational cloud.

For example, as Bratton writes:

"The stack's visual geography amplifies economies of mutual simulation between land, image and interface by redefining the surface of the Earth as a living and governable epidermis, and recomposing that skin as a bio-informational matrix enrolled into other hard and soft systems. As a landscaping machine, The Stack combs and twists settled areas into freshly churned ground, enumerating input and output points and rerendering them as glassy planes of pure logistics. It wraps the globe in wires, making it into a knotty, incomplete ball of glass and copper twine, and also activates the electromagnetic spectrum overhead as another drawing medium making it visible and interactive, lining the sky with colorful blinking aeroglyphs. The Stack walls off whole layers of that spectrum for private purposes by optimizing it through finer and finer atmospheric grids, turning location geolocation into application engineering. Its image of infrastructure and the infrastructure of the image flip-flop their respective works, repositioning geoscopy as

105

geoaesthetics and geo aesthetics as geoeconomics."

One early pilot project called Rainforest Skin, looked at the quantity of carbon contained within the planet's rainforests. The project combined data sets from satellites, unmanned aerial vehicles, and ground-based sensor networks to estimate the forest's carbon stock and flow. This could allow for trading and risk management.

The energy consumption of a planetary computational infrastructure can be integrated into sustainable energy infrastructures. One couldn't run a planetary computational infrastructure off fossil fuels. Yet with a specificity of use that is necessary in areas that require such platform integration the use would be critical in creating a holistic solution that is in the end more sustainable and ecologically integrated.

E7 Architecture Studio, where my father and I work together as fellow directors, has developed a proposal for a platform architecture, an interface of cooperating softwares that give organization to informational matrices in model format that provides the integration of ecological components and systems as correlate to an object model with facets of layering attributes and outputs of charts and graphs that give statistical and quantifiable metrics of ecological phenomena, as demographics, changes of population and movement of energy. The object model functions through cross-references as an open geospatial correspondence where multiple connected agencies co-author the multilayered infrastructures of the built environment. The model is open sourced and referenced through educational programs whose curriculum outlines ecological interdependence of complex systems.

Systems like these can be used in Virtual Community Action Planning (VCAP), a dynamic and innovative process that combines the deconstruction of a city with the integration of the Digital Gaia API, fostering ecological interventions and community transformation. Through the creation of 3D virtual context models and embedded location mapping systems, VCAP enables the identification of suitable areas for project interventions and impact investments. By leveraging the power of digital simulations and the ecological capabilities of the Digital Gaia API, VCAP facilitates the measurement and confirmation of the efficacy of ecological interventions, providing a tangible and measurable impact on the environment.

The first step in the VCAP process is the creation of 3D virtual context models of the city. This involves the use of advanced technologies such as aerial imagery, LiDAR scanning, and satellite data to generate highly detailed and accurate representations of the urban environment. These virtual models serve as the foundation for the subsequent stages of the process, enabling stakeholders to explore and interact with the city's digital replica.

Embedded within these virtual context models are location mapping systems that allow for the identification and categorization of areas suitable for project interventions and impact investments. These mapping systems utilize geospatial data to assess various factors, including ecological conditions, infrastructure availability, and community needs. By analyzing this information, VCAP enables stakeholders to make informed decisions regarding the most suitable locations for implementing ecological projects.

The integration of the Digital Gaia API plays a crucial role in the VCAP process. The Digital Gaia API is a powerful tool that simulates and models ecological systems, allowing for the evaluation of the potential impacts of proposed interventions. This integration empowers stakeholders to assess the effectiveness of ecological projects in a virtual environment before implementing them in the physical world. By leveraging the Digital Gaia API, VCAP provides a means of confirming the efficacy of interventions and optimizing the allocation of resources, leading to more sustainable and impactful outcomes.

Crucially, VCAP is designed to engage the community throughout the entire process. The virtual context models and interactive platforms enable community members to actively participate in shaping the transformation of their neighborhoods. Through virtual engagement tools, residents can explore proposed interventions, provide feedback, and collaborate with stakeholders to refine and improve project plans. This inclusive approach ensures that the community's voice is heard and that the interventions align with their needs and aspirations.

By deconstructing the traditional approach to urban planning and integrating the Digital Gaia API, VCAP revolutionizes the way communities and stakeholders can envision and implement ecological interventions. The combination of 3D virtual context models, location mapping systems, and the powerful simulation capabilities of the Digital Gaia API creates a comprehensive framework for sustainable and impactful community development. By enabling stakeholders to visualize, measure, and confirm the effectiveness of ecological interventions, VCAP empowers communities to transform their neighborhoods, fostering a more sustainable and resilient future. The maintenance of our environment requires that we have platforms that are inclusive and embedded with educational and participatory infrastructures in order that we can all participate in the care and visualization of a better future for our environments.

VCAP is a curriculum process, the following is a walk through the design and analytics process that Virtual Community Action Planning outlines. This process is refined and developed over time to better integrate participants and desired results.

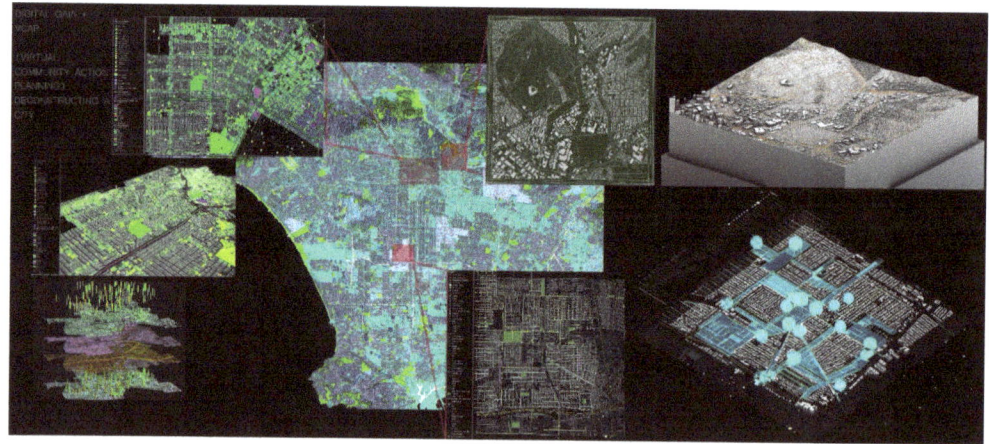

Phase 1: Curriculum and Education

Week 1: Measurement

Since the objective of improving our community is based on numeric value it is necessary to know what are the values that we should be measuring in our community. The initial one hour lecture will cover what is consistently measured in a community in order to define a criteria of desirability and what is used to identify a community as undesirable. How we use specific tools to measure these relationships will be explored in week one. There will be examples of basic GIS functions: Search, Buffer, and Sort. Exposure to the three or four web applications being used to define neighborhoods will be demonstrated and discussion relating to their backend data structure will be explained. Terms like, Symbology, Topology, Network will be demonstrated during week one demonstration lecture.

Week 2: Utility Systems

One-hour discussion and presentation relating to major systems necessary to support neighborhoods. How is this basic infrastructure support network taxed or how could it be augmented to support a better neighborhood. Examples include looking at energy, water, sanitation, and stormwater.

Understanding and documenting how these infrastructure systems are managed and designed to be dependent on a real time model. Each of these primary utility systems will become more regionalized in most areas and with that will be opportunities for innovation looking at how these systems interconnect.

Week 3: Main Building Facility Types

During week three, the main ideas of using a 3D modeling tool are covered. Using Sketchup and KML files for geo location are explored. How to augment a neighborhood using libraries, and build a collective model with specific building types are explored.

Example: Hospital Building

The instruction for using Sketchup will allow the sharing of these building types. This will allow for group model building.

Week 4: Community Activity

Week four includes learning, making, purchasing, visiting, worshiping, shopping, repairing, studying, working, mentoring, gardening, recreating. Students will prepare a 10-image PowerPoint presentation about the human qualities needed to conduct each of these community activities. Define the method to assess how a neighborhood can capture these specific types of assets. How is a figure ground relationship developed? How is the scale of the neighborhood experienced? Line of sight, smell, physical experience?

Week 5: Natural Systems

The intent of the week five investigation is to show the numerous natural systems that extend across neighborhoods that should be digitally documented. This system is the underpinning concept for Week 6: System Dynamics. Through this exploration, qualities of neighborhoods and their direct interconnection with natural systems will be identified.

Week 6: System Dynamics

System dynamics is a computer-aided approach to policy analysis and design. It applies to dynamic problems arising in complex social, managerial, economic, or ecological systems—literally any dynamic systems characterized by interdependence, mutual interaction, information feedback, and circular causality.

We:

- define problems dynamically, in terms of graphs over time.
- strive for an endogenous, behavioral view of the significant dynamics of a system, a focus inward on the characteristics of a system that themselves generate or exacerbate the perceived problem.
- think of all concepts in the real system as continuous quantities interconnected in loops of information feedback and circular causality.

Then we formulate a behavioral model capable of reproducing, by itself, the dynamic problem of concern. The model is usually a computer simulation model expressed in nonlinear equations, but is occasionally left unquantified as a diagram capturing the stock-and-flow/causal feedback structure of the system.

Week 7: Blog Technology

The intent of week seven is to prepare the presentation and pull together the investigation conducted during the six previous weeks. There are several tools that can be implemented to prepare the presentation.

Intent of the presentation is to show how existing data and constructed data are merged together. This is often titled Mashups. These Mashups will document community-based observations regarding specific Community Action Proposals.

Week 8: Final Presentation

During week eight each student will have completed their blog for Community Action Proposal. This blog will be the launch point for a process to activate the actual World Peace One project. Through a series of conversations the blog will identify a collective larger community project and how "The keys to the success of group modeling building efforts appear to be engaging stakeholders, sharing mental models formally, assembling and managing complexity, using simulation to test scenarios and support or refute hypothesis, working toward alignment, and empowering people to have confidence in the strategies that emerge."

Phase 2: Implementation

First Job: Existing Project Identification, Modeling, and Integration

- Bring information from 25 Projects and put them into the Virtual Community platform model
- Invite the 25 project managers into the laboratory and ask them what they are doing and what they need. Each of the 12 youth employees are assigned to various agencies.
  - Youth employees can offer proposals to support these 25 projects.

The goal is to integrate and communicate between the projects to have a connected solution and holistic vision of solutions that address the interdependencies of the environment and the connectedness of the community. Using Watts Rising as a Case Study with multiple simultaneous projects. This diagram correlates topics and categories of urban interventions with professional sectors and their project locations.

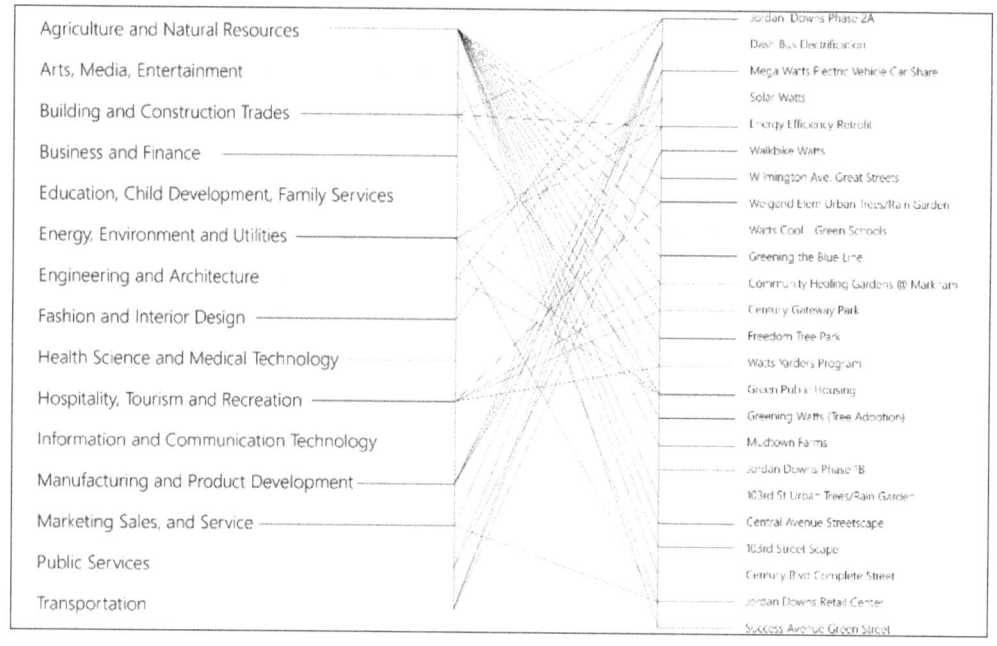

Agriculture and Natural Resources	Jordan Downs Phase 2A
Arts, Media, Entertainment	Dash Bus Electrification
Building and Construction Trades	Mega Watts Electric Vehicle Car Share
Business and Finance	Solar Watts
Education, Child Development, Family Services	Energy Efficiency Retrofit
Energy, Environment and Utilities	Walkbike Watts
Engineering and Architecture	Wilmington Ave. Great Streets
Fashion and Interior Design	Weigand Elem Urban Trees/Rain Garden
Health Science and Medical Technology	Watts Cool - Green Schools
Hospitality, Tourism and Recreation	Greening the Blue Line
Information and Communication Technology	Community Healing Gardens @ Markham
Manufacturing and Product Development	Century Gateway Park
Marketing Sales, and Service	Freedom Tree Park
Public Services	Watts Yarders Program
Transportation	Green Public Housing
	Greening Watts (Tree Adoption)
	Midtown Farms
	Jordan Downs Phase 1B
	103rd St Urban Trees/Rain Garden
	Central Avenue Streetscape
	103rd Street Scape
	Century Blvd Complete Street
	Jordan Downs Retail Center
	Success Avenue Green Street

Second Job: Neighborhood Council, Community Engagement

- Train Neighborhood Council Leadership to be familiar with VCAP process & model
- Watt's Neighborhood council has approx 15 people
- Student is the facilitator of councilmembers connection to VCAP

Third Job: High Level Design Solutions

The following is an Engagement Process as a 7-Step Workflow integrating Contextual Analysis of VCAP With High Level Design and Investment Strategies. Facilitate and carry out conversation with high level designers to implement design interventions and solutions.

.

1. GIS Data Composition and Organization (Analysis)

The service provided during this phase will be an assembly and organization, perhaps even creation, of a system that gives information of the built environment as a series of infrastructural, social, and ecological systems. These systems are measured statistically and spatially in their dimensionality. Informations will exist as cartographies, symbologies, topologies, and networks, etc. Systems represented include utility systems, main building facilities, community activities, natural systems, etc.

In the data composition process, context is generated and relevance gives value to future design initiatives. Programmatic proposals can respond to data narratives and context that give relevance of use to the community. Data is collected and grouped into meaningful categories that exemplify a co-functionality of the built environment systems.

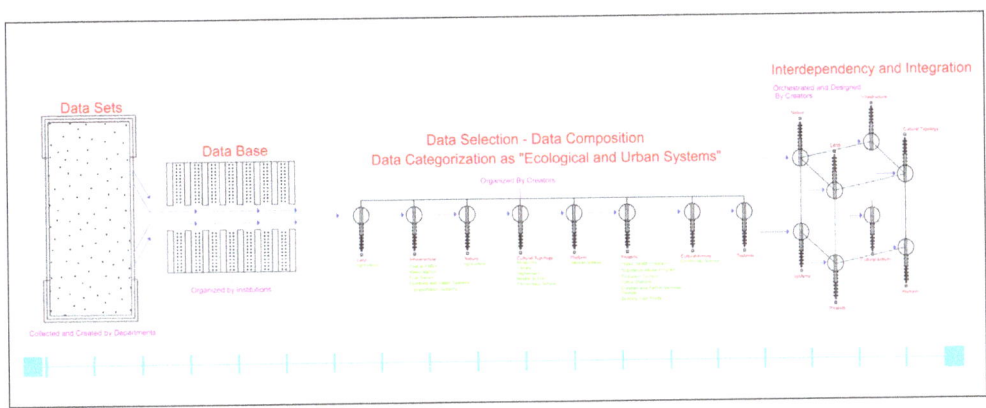

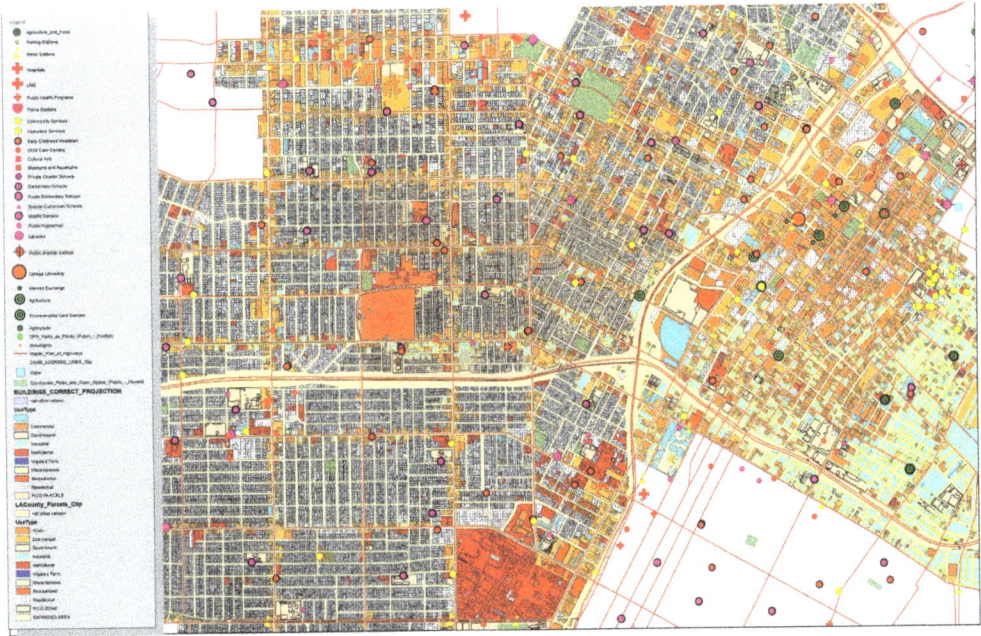

2. 3 Dimensional Contextual Model (Site Generation) (Analysis)

3D context models contain imported data from the 2D GIS data composition. Generating the 3D model creates a model space that is **113** more engaging for future design initiatives as BIM (Building Information Modeling) becomes a preferred method of designing sophisticated urban solutions.

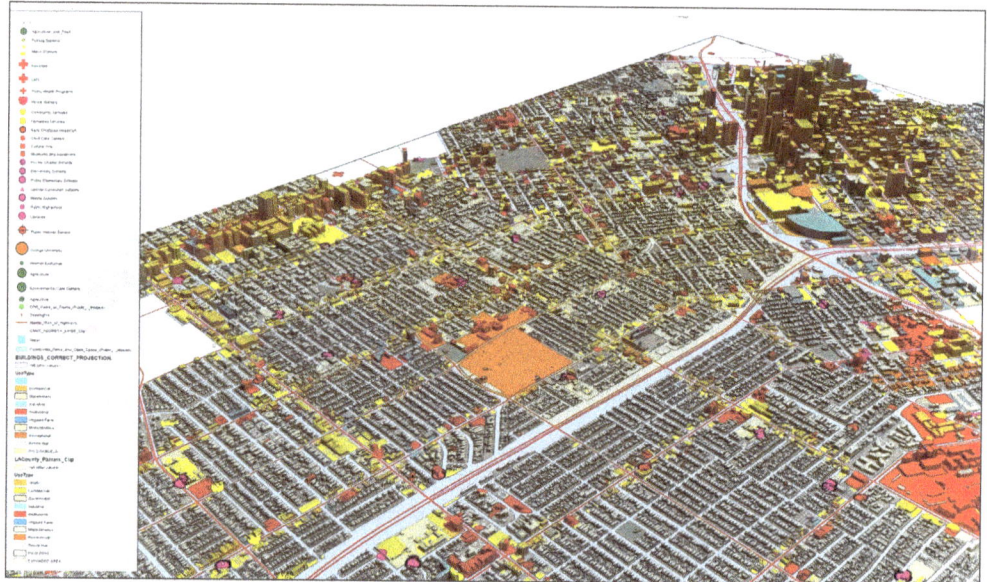

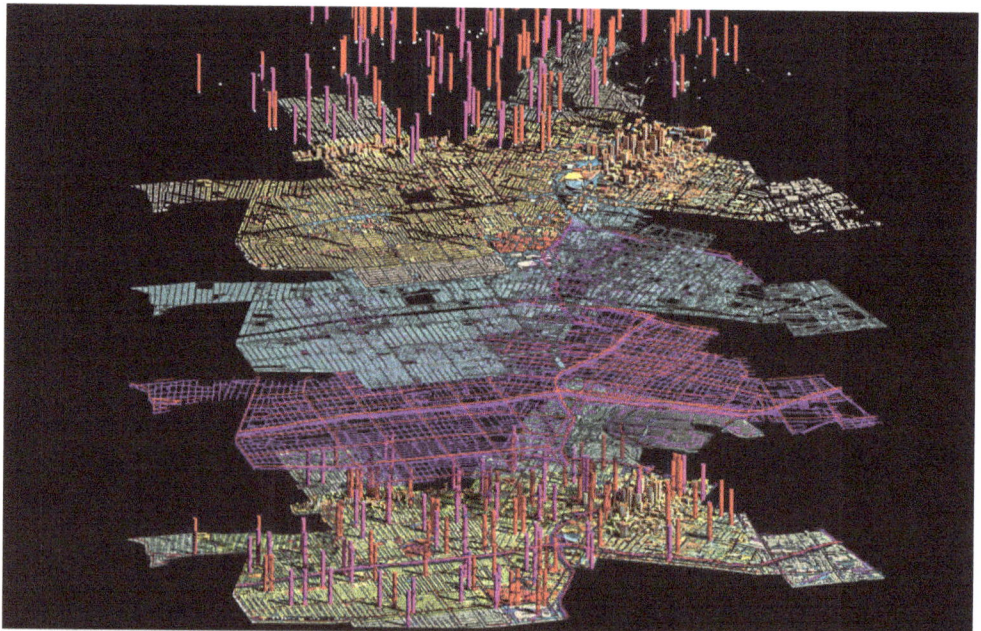

3. 3D Model Zoning Scheme + Large Scale Schematic Design (Analysis and Design)

Responding to a context of data analysis that describes system dynamics as existing social, ecological, and infrastructural systems, a zoning strategy as a large-scale schematic design begins to articulate new impositions and augmentations to the built environment that intersect, interact, and relate to existing dynamic movements of energetic systems. Cultural institutions and green spaces can be linked through schematic transportation connections. Methods of urban agriculture and food distribution may be rethought. The 3D model from step 2 is available for manipulation and community members may discuss radical transformations with agencies capable of carrying out such changes. The map below displays points of community engagement as gathering spaces and community components that are nodes on a distributed network, the points forming a social infrastructure in the neighborhood. The social infrastructure is linked by green infrastructure that nourishes biodiversity and brings health and wellness.

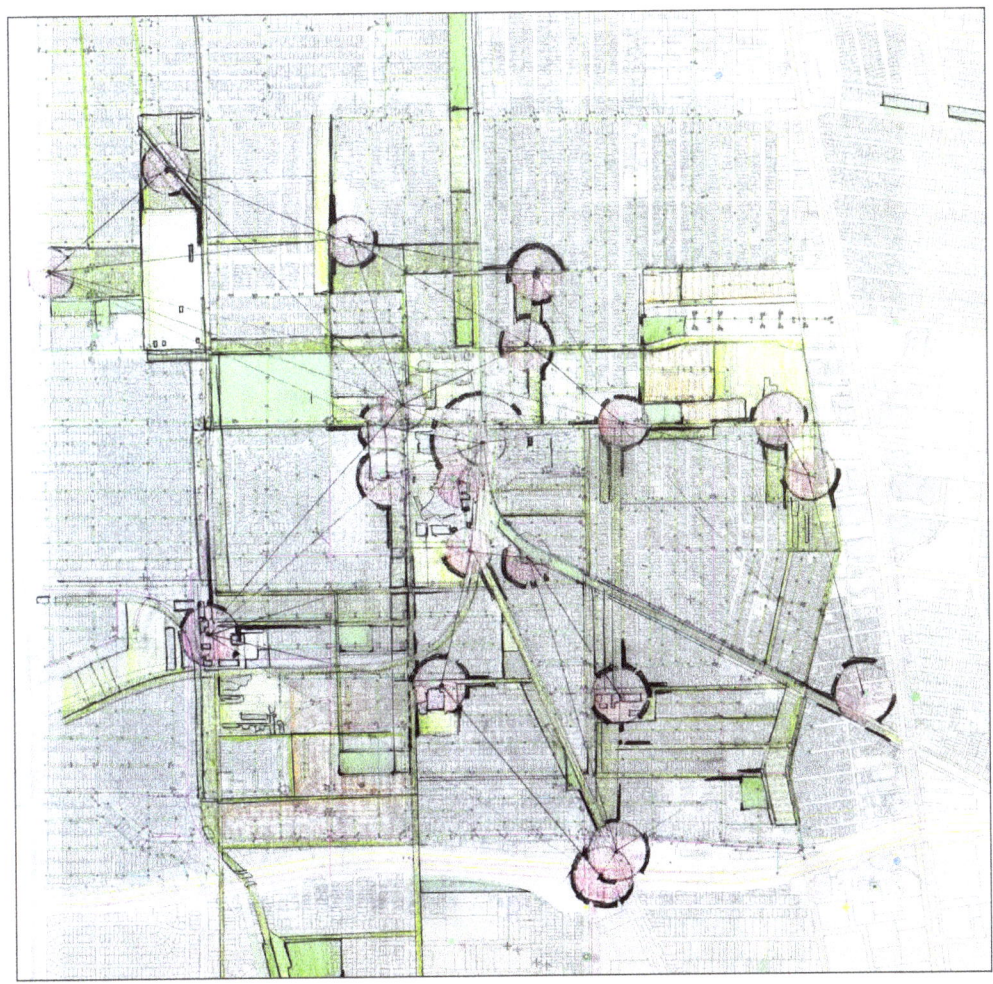

4. 3D Model - Parametric Organizational Scheme (Design)

Parametric Organizational Schematics add an additional layer of organizational matrices as grids and geometric striations that give holistic compositional reference. Geometric armatures or outlines provide context for infrastructures and architectural projects to follow as they are integrated into a contextual system, unifying and creating harmony amongst multiple projects and systems.

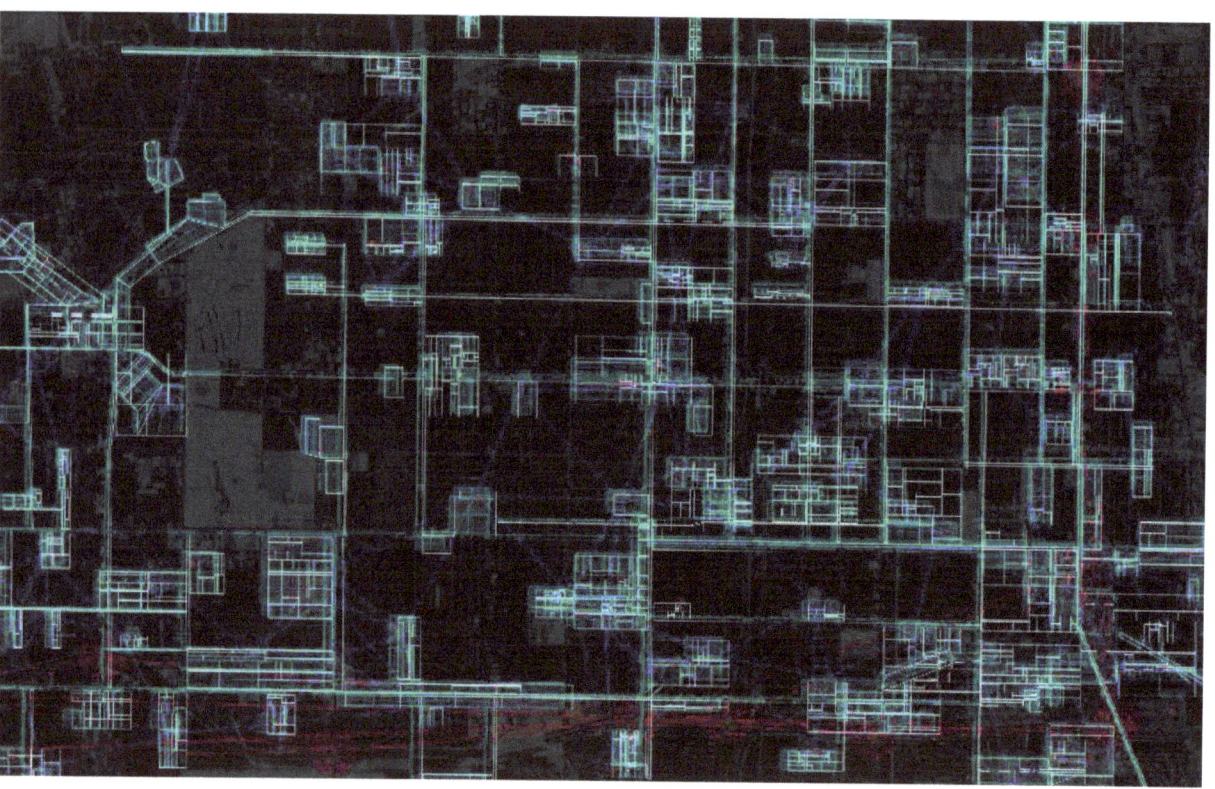

5. Architectural Design Service

Architectural design here is contextualized by the previous data sets and uses a parametric armature to ensure a cohesion between the multiple elements involved in the architecture.

6. Design Build Development (Product)

Following various forms of schematic design integrated plans are created for correspondent initiatives for multiple projects. During a design build process the environment is built and actuated. We move from data-rich models with schematics into built form, a planning process with phases. Within a targeted area, various projects are initiated within the scope of an overarching vision of an environment. This way, reforestation and urban living can be calibrated as a co-functional system design. Biodiversity preservation, recreation, and education can be sought after with parallel design initiatives. The interdependent nature of our environments that exists within corresponding components can be visualized and then delivered in this step. Following construction, the feedback loop is established when the multiple layers of real built projects are uploaded in a virtual model in step 7. The built environment is iterated upon with models of itself that correspond to each other.

7. Virtual Environment Simulation (Presentation)

Digital Twin of London for planning and architecture. Image from Creative Commons

7-Step Process Results in a Built Environment Solution

Integrating the open sourcing of the model into an educational program ensures mass labor required to generate simulation granularity and fidelity to achieve ecological reflections of causal scales. Contextual map sites and urban fabrics become engendered in cultural communication at the education level and vertically integrated into public and private industry through internships. The model culture of simulation of the environment becomes an open social exchange of how the environment can be augmented. Illustrations, interventions and installations of the social fabric, to be more suitable to the vitality of ecological needs, may be implemented or envisioned through the simulation models. The isomorphic correspondence of the internet and the urban fabric as bi-product entities requires this cooperation.

The ecological project executed with communal engagement, interdisciplinary coordination and virtual model coordination is an oriented operation system towards a continuous ecologically harmonious system. The platform as a shared representation of ecology is accessed in a high fidelity model format that organizes accessible layers of the built environment and their interdependent co-functionalities. The platform ensures co-functionality of the layers through an interdependent criteria represented in the structure of a cube and its eight vertices: Land, Nature, Cultural Activity, Cultural Typology, Systems, Projects, and the Platform itself. The platform is accessed through representations that reveal its components, through cartographies, drawings, and diagrams that articulate the functioning elements of ecological processes. Project agency, square footage, area of influence, and performative metrics are accessed through the layered representations. The fidelity of the model could be accessed in simple diagrams or highly articulated Building Information Model Simulations that explain deeper functionality.

Instrumentalization is defined as the encounter between the community and professionals with the platform. Instrumentalization of the platform may occur in different spaces throughout the community, such as schools, office space, public space, galleries, institutions, households, virtual interfaces, and collaboratories. Instrumentalization of the platform is a cultural activity towards a collective vision of ecological harmony.

Methods of modeling, drawing, and mapping the built environment and biosphere each reveal constituent performances and tendencies within the respective systems. As descriptive and analytical languages, methods of representation describe phenomena. Localized entities, defined by agency or formal determination as buildings or continuums of infrastructure, manifest with governing principles that are fundamental and inherent to their ability to construct themselves. The principles of construction include geometric specificities of structure or a distribution of forces that are residual to the system and its articulate language. The forms are self-referential and exist in independent formulas that are contextualized. The descriptions of context as composed boundaries with attributes and components creates an opportunity for dialogue and interaction with self referential entities that engage a context. Referential phases allow one to trace the creative architectures of localized entities and their causal complexity as connected with a complex context.

CHAPTER 9

Global Terrestrial Reality Case Studies: The Resource Extraction, Distribution, and Consumption Dynamic At the Planetary Scale—The Colorado River Basin

The research of this chapter was conducted by the guidance of Lauren Bon and the Metabolic Studio with whom I am grateful for the lessons learned.

At many locations on the planet, including Africa, the Amazon Rainforest, and the United States, extreme amounts of Earth resource extraction and processing is taking place. Materials for iPhones and vehicles and construction are being extracted from the Earth and traded around the world in major cities that consume and use the resources for market purposes and lifestyles.

Resources are extracted in mines, oil fields, oil wells, quarries, fishing locations, agricultural sites, deforested forests like the Amazon itself. The resources extracted from the Earth are processed in robust energy hungry refineries and factories, there the resources extracted from the Earth are transfigured to be used for market distribution, often reconfigured in manufacturing processes to create consumable materials, like metals, woods glass and concrete, perhaps computer chips or asphalt, or vegetables, fruits and meats.

Our dense and populated megalopolises are consuming at rates that encourage resource extractions. The metropolis way of life, with its rate of consumption and production, creates a demand for imported materials and resources from around the world. A global infrastructure of sea routes and ports, roadways, highways, and railroads has accelerated the rates in which resources can be exchanged and consumed globally.

The megalopolises themselves and their more local infrastructures are also putting pressure on their immediate environment. The world's largest metropolises like Tokyo, Los Angeles, and New York are straining not only their regional environments but also the environments that are supplying their consumption rates. On a globalized planet, we have many cities that are consuming resources on a planetary scale. The list goes on and on: London, Paris, Cairo, Beijing, Shanghai, Dubai, etc. Sea routes and land transportation routes create a global network for resource extraction and consumption with the primary nodes being major cities and the locations of the resources.

Are we aware of our rates of extraction? Are we aware of the pressure our extraction rates put on the environments from which we are extracting? Are we aware of the rates of consumption of our major cities? Are our infrastructures for water distribution, food production, transportation, ecological preservation and cultivation, education, and housing going to hold up for the rest of the century and for centuries to come? If we continue business as usual and consider the current methods of extraction and consumption as normal, the answer is likely no. We may face a global civilization collapse because at a certain point our dead-end waste cycle that does not recycle our consumables is going to leave us with nothing left to consume.

The principle of sustainability is that we can sustain our future, likely accomplished with temperance, moderation. The biosphere retains a balance of consumption and energy exchange in a planetary homeostasis. There is an actual system in which the planet has arrived at abundance and resource accumulation through environmental processes that function on a certain scale and duration of time with certain rhythmic cyclical intervals. The metabolic clock cycle of the Earth determines the seasons, and over centuries, the creation of great plains and ecosystems as tectonic plates shift and glaciers melt. This is glacial time. There is a planetary reality to the systems of life sustenance and our major cities and global infrastructures must acknowledge these realities for the ultimate flourishing of humans and all life on Earth.

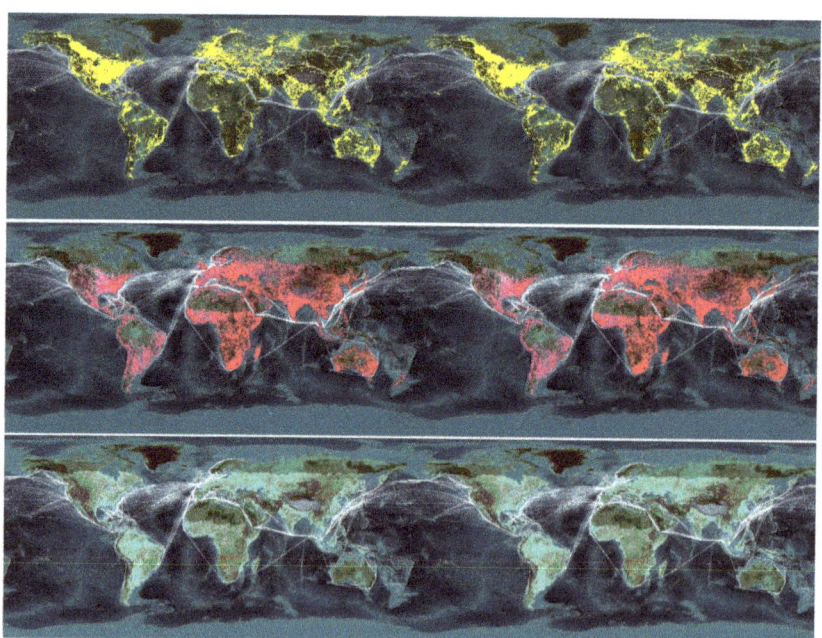

The global cartography is arranged to center on the Pacific Ocean, where a significant amount of trade routes are being continuously used. The series of maps seeks to visualize the planetary scales of resource consumption and exchange. In yellow we see the locations of resource consumption. The white arcs over the sea are trade routes that are distributing the resources globally. The red network of lines is the land transportation system of resources. The images together illustrate the planetary scale of resource extraction, exchange, and consumption in a series of flowing vector graphics. The Earth is a continuous ecology that circumscribes the planet with interdependencies networked in its totality. The crust is a continuous membrane, in some areas submerged under oceans and in some areas rising above clouds. The terrestrial reality is ancient and eternal as the universe, as the Earth is a cosmic phenomena. Our humanly contrived political, economic, and engineering solutions seek to domesticate the complex reality of a terrestrial system; yet we might consider a very deep humility in the face of such immense powers that all of us depend on for life.

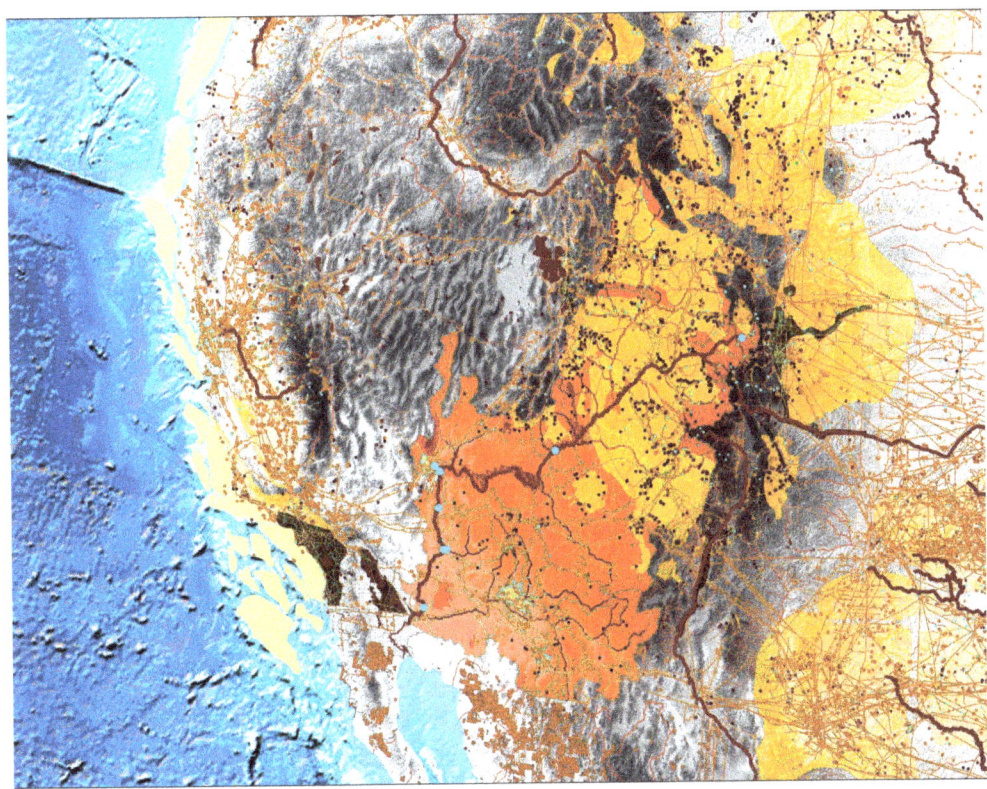

This geospatial cartography model of the Colorado River Basin identifies interdependencies in the life-web of the biosphere and infrastructure at the scale of 250,000 square miles. The Colorado River supplies drinking water for 40 million people and irrigates five million acres of farmland. What happens when it dries up? Both in Los Angeles and Denver, 50% of the water is from the Colorado River. Denver water serves 1.4 million people. Los Angeles uses 524 million gallons of water a day. The dams power 780,000 homes for 1.3 million people with hydroelectricity. Near the delta of the Colorado River is the Imperial Valley with 460,000 acres of farmland, which is $2 billion worth of agriculture capable of making 161,000,000,000 small healthy meals. Next to the Imperial Valley is Salton Sea where there are 32 million metric tons of lithium. The electric car industry is swarming this area to mine the lithium, like it is a gold rush, to make batteries.

Geochemistry or the chemistry of the Earth is critical for terrestrial architecture. Our megastructure megalopolises need to be reconsidered as highly interwoven with our terrestrial systems. The biosphere has methods of handling homeostasis, equilibrium, and sustainability. Are we conscious of these dynamic realities of the biosphere?

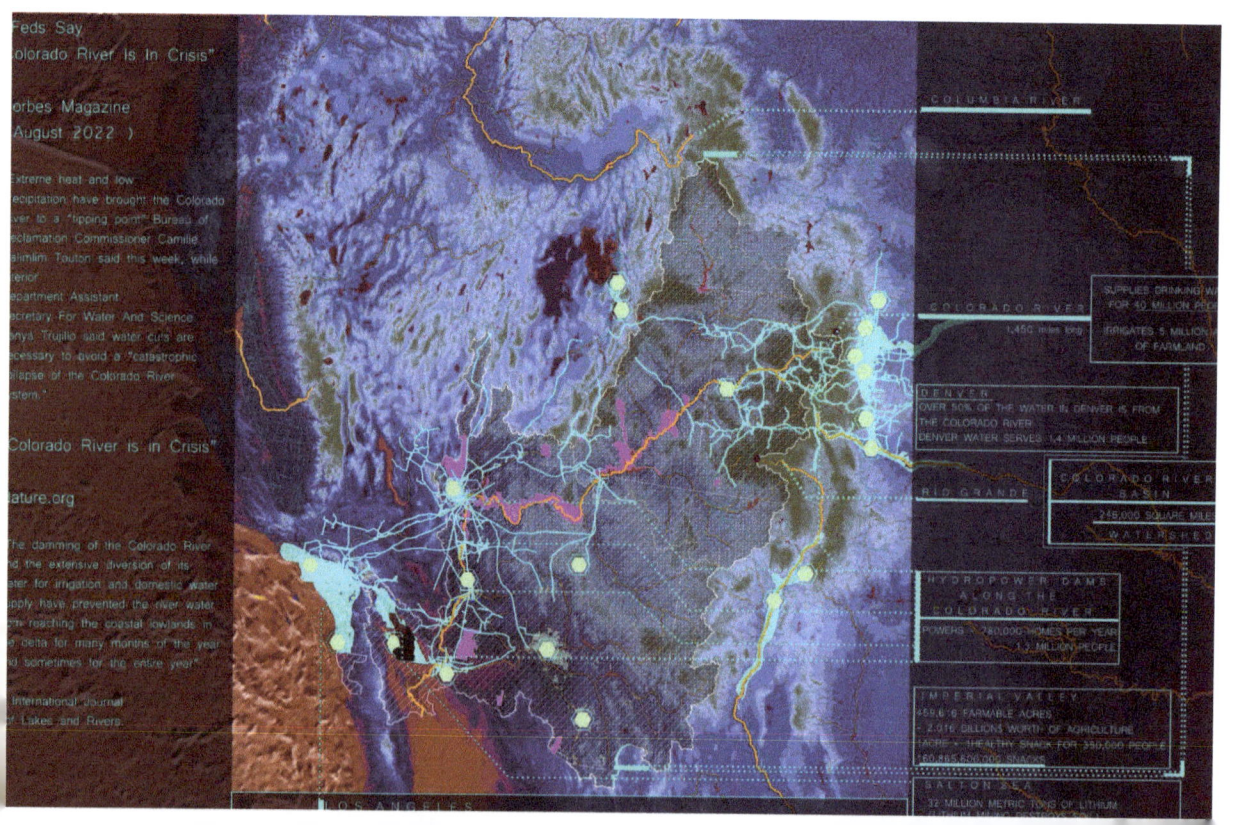

This map of the Colorado River system that we constructed with Lauren Bon and the Metabolic Studio diagrams show how the infrastructure of the river links across multiple cities, serving multiple cities with water and electricity. From this map we are able to see how Denver, Los Angeles, Phoenix, and other cities in the intermountain west are linked via the watershed infrastructure of the Colorado River. The green dots are cities that are infrastructurally connected to the River while the cyan splotches show the populated areas.

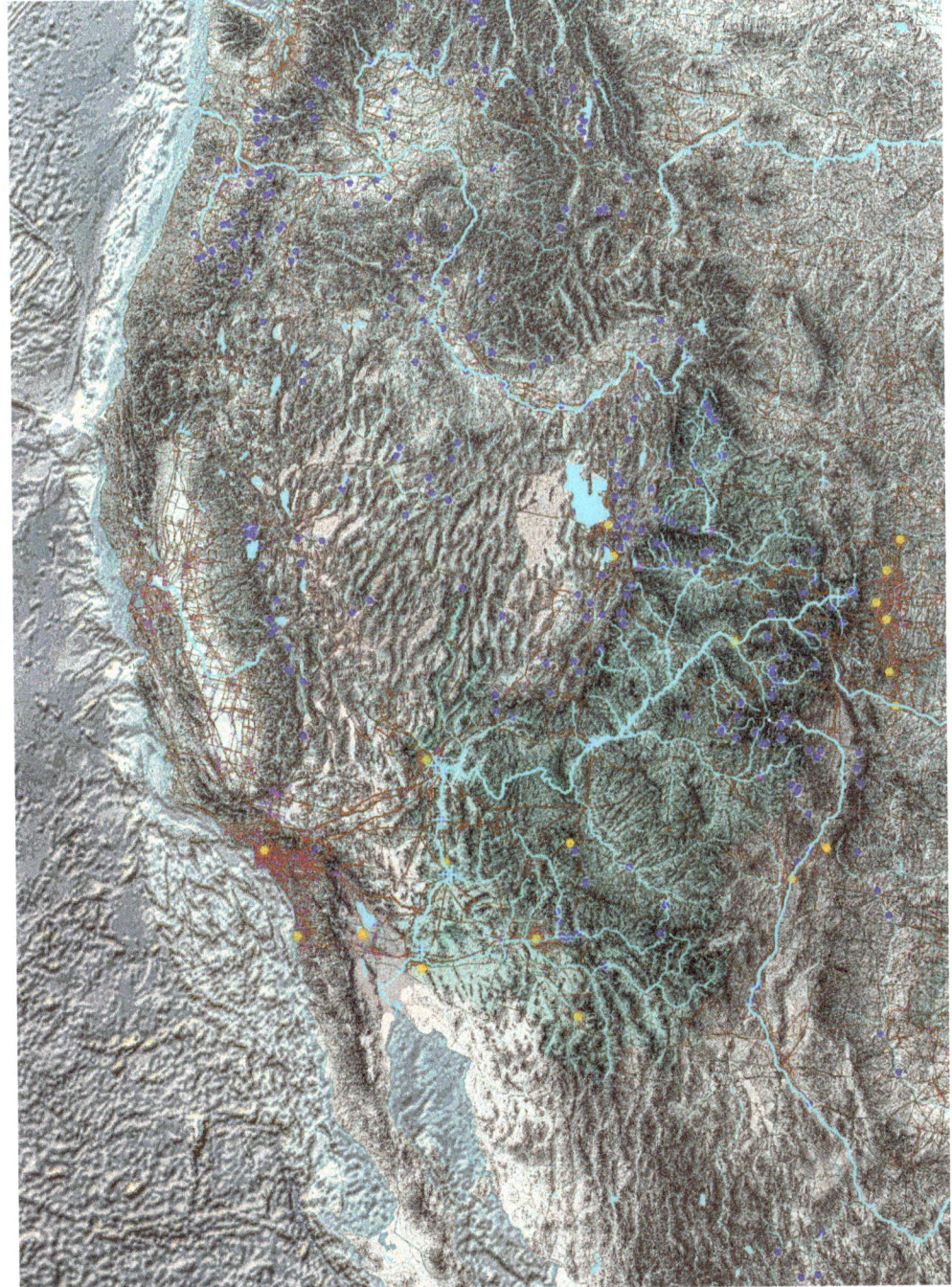

In this other cartography constructed with Lauren Bon and the Metabolic Studio, we see a slightly different approach to the representation with greater emphasis on the topography of the landscape as the water has carved out ravines, mountains, and canyons on the continent. At Metabolic we discussed the analogy of the heart, the rivers being like arteries. The cities linked to the Colorado River via infrastructure are marked in orange.

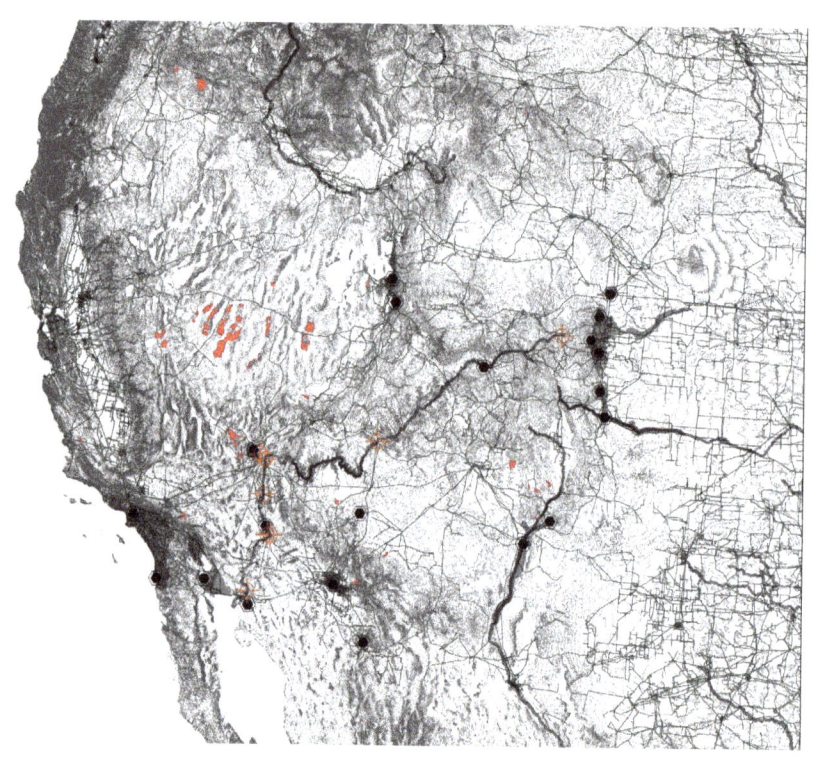

Another Metabolic Studio Map looks at the populated areas, infrastructures, and the river in a heavy black. The cybernetic nature of the watershed becomes very present as the two systems read in simultaneity. We also examined how horses are able to locate water and where they gather marked in red. Understanding how biodiversity lives within the landscape is a powerful observation that informs and educates us to see what are the advantages and available sustainable ways of living that are present on the land. A regional cartography of the Colorado River is a case study for ecosystems around the planet to be mapped, modeled, and understood in greater fidelity.

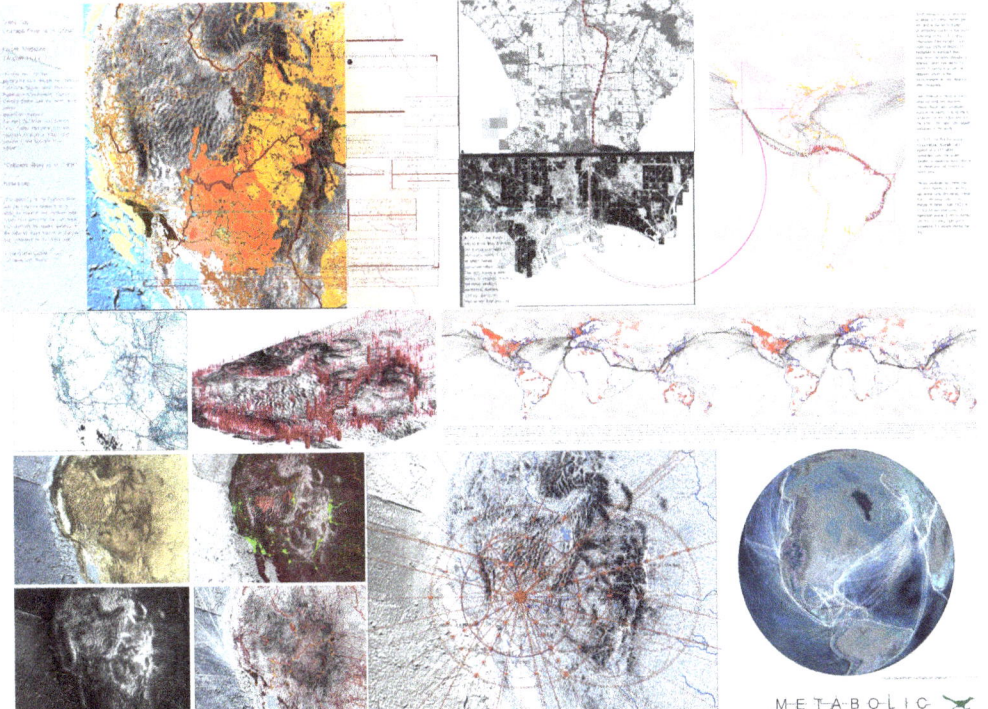

This image that we constructed at Metabolic Studio illustrates how multiple maps and models link to each other. The global trade route map plugs into the local and broader ecology of Los Angeles, which is depicted in multiple map iterations. This linking between scales and information is crucial for a global terrestrial architecture. It is important we visualize how our ecologies are inextricably linked. A multi-scalar solution as a stack or network of models nested within each other creates a digital twin of the planet from which we would be able to model terrestrial transformations in ecological harmony. A platform would be designed to manage complex data sets and the interdependencies of these models. This layout displays several models in adjacency and their linked truth.

A paradigm of design and participation is possible where our ecosystems are mapped and modeled in highly integrated platforms in order that we may be able to design on the planet in a more harmonic way. Multiple models and maps can be coordinated within platform architectures that describe their coevolution, their co-functionality.

127

CHAPTER 10

A.T.L.A.S.: Apparatus for Technical Logic in Analytic Structure

The A.T.L.A.S. project was my undergraduate thesis at SCI-Arc. This chapter and the following chapter articulate two thesis projects and how we might approach a platform architecture for Earth Analysis and Stewardship. A.T.L.A.S. has an ability to organize complex compositions of information in a potential for spatially orienting interdependencies of information. The platform is a spatialized infrastructure of information in order to derive holistic solutions of synergy and synthesis to the built environment; practices of informational reference and contextual relativity must be developed. The built environment is composed of layered systems of biospheric activity, biogeochemical activity, energy infrastructures, socio ecological cultural systems, living fabrics, and economic scenarios, etc. Such informational complexity is fruitful for projects to achieve an interconnectivity within their context and simultaneously achieve a breadth of unity and inclusion when such systems are resolved in their interdependence. Such systems rely on information infrastructures for maintenance and their initial design genesis.

As an abstract geometry, this proposed platform architecture functions as an armature for conduits of information, simultaneously it is an armature for the energy infrastructures that sustain the energy required to manage information. Information travels through the structured aggregate in formations of circulation and adorned enclosures in fractalized multiplication, as in crystalline growth, with adjacencies and parallels indexing groupings of information as relative and codependent constructions. The information is to be stored and preserved as in archives and libraries and also projected, accelerated, proliferated, as in laboratories, creative spaces, and presentation spaces. Assembled in an integrated framework machina, a composition of frames, as rooms, screens, partitions, and levels, in sequence, articulate the details of information segmentation and stratification. Cumulatively they generate the territorialization and deterritorialization in

129

informational dependencies.

Structural additions, subtractions, or assemblages are resultant—projected by cumulative informational component aggregations along with physical infrastructural necessities of energy dynamics including thermal difference, light manipulation, and wind orchestration. Biological tissues, amber, and enzymes line the physical infrastructural hardware where combined technology intersects, retaining a clean and operable state.

The procedure of articulated perspectival histories within projected futures is referential to origins and interpolated future movements of information. The postulate projection manifests as a consideration between potential influences, variables, forces, and agents in their composition of generating a projected future architecture. The directional forces of expanding aggregates of knowledge meet counterparts that are, in turn, projecting their own emergent structures causing convergence and divergence, stratification, or supplementation around new centers—generating temporal knots that include various lineages of multiple contexts and histories that give multi-layered descriptions of the built-environment's fabric.

The platform is a strategy as interface to use spatial composition and spatial organization along with a multitude assemblage of organized digital display types to provide interaction between lexicological, graphic, institutional, historical, philosophical, and other infrastructures of creativity: museums, archives, libraries, documentation spaces, studios, spaces for inquiry and presentation, laboratories and decoding centers, illumination and conservation efforts. Jurisdiction over creativity is organized through curated identities that have the capacity for specialty over creative and analytic processes.

The machina configures human beings in an infrastructural and urban scale to be seamlessly articulated with intelligent machines that participate in the Apparatus for Technical Logic in Analytic Structures. It is a method of dialogue with forms of governance—postulating architectural environments for tools that are required for mega scale design and analysis operations, such as artificial intelligence, virtual reality, augmented reality, drawings, books, a multitude of softwares and hardwares—the collaboration between architects, philosophers, scientists, technologists, engineers, artists, mathematicians, and politicians with these intelligent machine entities and modes of information in these environments that curate dependency models.

Aerial Drone photography and rendering of the infrastructural project A.T.L.A.S captured along the Los Angeles River.

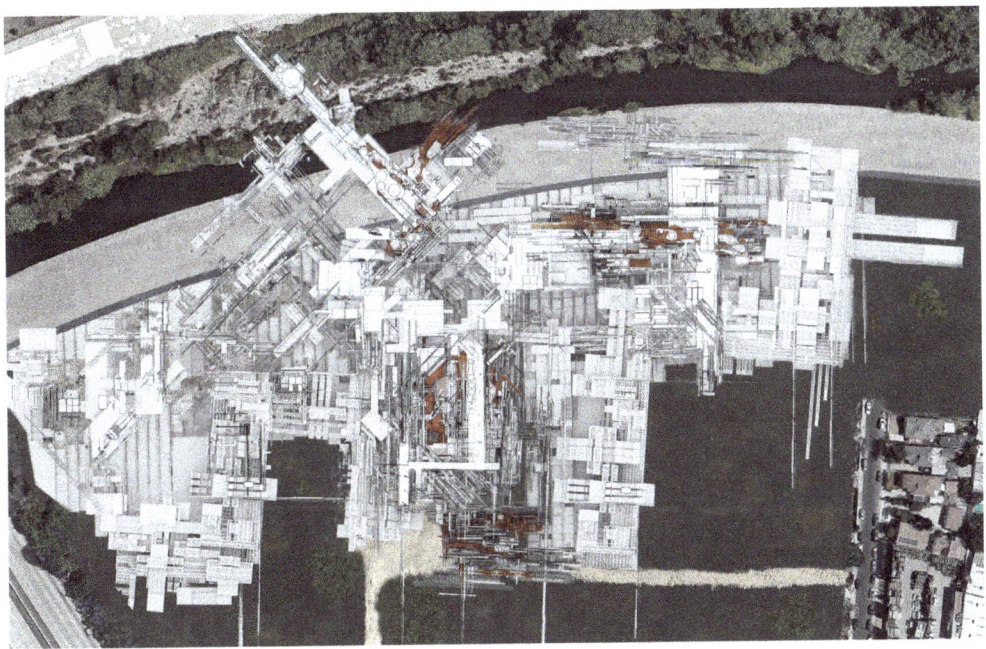

Floor plan shows shifting room spaces organized in peripherals, adjacencies, and sequences as a curated space of data contents like a sprawling gallery or library that is also an energy infrastructure.

A vantage point nested in the A.T.L.A.S. structure where one is able to see the multiple axes of cellular diffusion. The orange amber resin containing computer hardware is encased in its mesh as a preservative of the technology.

A cross section of the A.T.L.A.S. structure showing a core tower and a terracing array of platforms in blue.

CHAPTER 11

Instrument of Terrestrial Transformation

Generally, Instruments of Terrestrial Transformation (ITT) is about libraries of representation and description of the Earth as the foundation for projecting futures of the Earth. It is about the design of environments through the building as an instrument, a device for modeling. The building as an instrument is an institutional foundation for a dialogue between the Earth and the Design Disciplines. The dialogue happens in a feedback loop as in a cybernetic infrastructure.The Earth is described by disciplinary languages inside the building, in forms of analysis and comprehension, thus enabling the ability to foretell or create the future of the Earth.The building mediates transitions from virtualities into actualities or disciplinary languages existing in virtual representations and models.

The building embodies this virtual space, consisting of abstractions, dreams, pasts, futures, and memories that are constructed as translations of a once actual, now, concrete present. In *Gilles Deleuze: An Apprentice in Philosophy*, Michael Hardt writes, "The possible is never real, even though it may be actual; however while the virtual may not be actual, it is nonetheless real. In other words, there are several contemporary (actual) possibilities of which may be realized in the future; in contrast, virtualities are always real (in the past, in memory) and may become actualized in the present."

In the dialogue between nature's environments and the design disciplines Bruno Latour, in *Facing Gaia*, offers two paradoxes: That Nature is not our construction and surpasses us infinitely, while society is our construction and immanent to our actions. However in moments of science, nature is our construction in the laboratories where we construct understandings, in other words immanent to our action, while Society may seem beyond our construction and surpass us infinitely.

An instrument of terrestrial transformation is proposed here as a building, a combination of libraries, projection spaces, creative spaces, analytic spaces and laboratories (spaces of multimedia). The multimedia program embodies a virtual space, liminal in analysis, understanding, creation as a dialogue between Earth and humankind.

135

It is important to articulate in the foretelling of futures, what relationship between architecture and ecology is being sought after. Our ITT would articulate the possibilities of a synthesis between nature and architecture at an environmental scale, similar to the Biomorphic Biosphere by Glen Howard Small, envisioned between 1969 and 1973. The term biomorphic evokes images of the biological organism and the term biosphere meaning the crust of the Earth.

In the Biomorphic Biosphere the term Megastructure is redefined as "a combination of many objects and systems in ways that complement each other and create a diversified but harmonious whole. Nature offers examples of huge ecological megastructures that combine mountains, valleys, forests, streams, animals, etc. Our problem as human beings is to interact with these natural systems in harmonious ways. The ultimate megastructure being the earth as a whole," according to Small's blog "Small at Large."

The idea of the Biomorphic Biosphere was to turn the city into an oasis, the structure a necklace of cruciform compression members linked at the tips by tension cables forming a stilt-like connection with the ground and an armature upon which natural systems would grow. Nature was here viewed as a technology. The systems of the human hand as a structure with skin was reconstructed as a structure overlaid with various functions including a skin. One could imagine that a structural armature could, in synthesis, provide infrastructural components for the diversity of ecological systems in both the natural and artificial world.

Foretelling of futures in a relationship between architecture and ecology conceptually is a sublime configuration, evoking a cohabitation with nature's most treasured or even sacred sites as Yosemite or Mount Whitney. Conceptually, the virtual and the actual in their interaction may be organized systematically based on components and their functions. Certain components could be simulation categories of environment, taxonomies, cyber-infrastructure composition, user-services and languages, epistemology, and simulation instruments.

Hans-Jorg Rheinberger states in "Gaston Bachelard and the Notion of Phenomenotechnique," when speaking about scientific instruments in general, that "The instrument sits in the center of the epistemic ensemble" as an organization of knowledge.

In the building as an Instrument of Terrestrial Transformation disciplines of knowledge are organized. The virtual abstract ideal geometries of mathematics and crystallography through construction geometries become Soufflot's Parisian Pantheon; similar proportional construction geometries become the master plans of Le Corbusier. Calculations of geometry triangulate landscapes into dimensional elevation maps and the details of a voxelized digital universe represents flows of energy within the Earth and built environment. Geometries and ideal languages materialize through conceptual processes. The process of an abstract ideal becoming concrete and present could be called a miracle.

The real world or actual world is bridged by conceptual worlds to a geospatial world of domain-specific languages, including the formal languages of mathematics and logic, geometric languages, georeferencing languages, and topologic languages, etc. The geospatial worlds participate in the description of the Earth.

As descriptive components in abstraction, instruments of simulation discretize and segment information to replicate or imitate a behavior process. Components of such instruments include lenses, trackers, detectors that enact a behavior and analyze its phases identifying causal relationships, conditions, mechanisms, rules, or structures. The virtual discretization and segmentation of information organizes ecology and the built environment within the building.

The building diagrammatically opens itself in branching pathways that bridge the interdisciplinary acts of description and foretelling. On one wing, the study of ecology is bifurcated into biology and ecosystems, hydrology and geology on the other wing the design disciplines are bifurcated into architecture, landscape, and urban design. The outer edges of each wing are allocated for information processing, in a spectrum from edge to more central in the order of memory systems, analytic and creative procedures. The central spaces are composed of projection spaces, where the creative visions foretelling the future are displayed in a communal setting.

The components of the program are represented by an array of images. Ecology, land resources and biology systems are cartographically described. Informational processing components are arrayed as servers and libraries in a networked composition. Projection spaces are composed of enveloping walls with projections that surround the occupant; also postulated are augmented reality environments where parallel physical virtual worlds coexist. Design disciplines are arrayed in studio spaces. Laboratories and various living spaces are also provided.

Informational matrices durable of continually increasing complexity while maintaining a disciplined structure are composed through studies of algorithmic proportions. Grids are projected in spatial ratios, intervals that are proportional, generating recursive relationships. The components of the grids are stratifications that stretch relatively vast distances while being subdivided in relative alignments.

An array or constellation of points with inherent organizational principles are projected into structures of organization forming an interdimensional space of expanding axes of mobility or freedom. Relational frameworks of lines are projected spatially within themselves. Perspectives are nested within perspectives, volumes within volumes with proliferating organization across layers.

The genesis of the form is imbued and encoded with organizational principles that discern its future assemblage as it is animated. A rhythmic organization of space segments and discretizing the variant disciplines. Bridges span the vast horizon of interdisciplinary knowledge, their intersections marking intervals and segmentations.

The dendritic passageways and walls of information display are navigated, toggling and switching between content in stacks and sequences. The building or instrument that facilitates the curation of visions is articulated through a network of pathways that exchange the transmittance of individuals and information. Bridges span connecting studio spaces to projection spaces; interval breaks in the spaces are disciplinary and organizational. The axes towards the center of the building are for precessional movements of knowledge and information into the central projection spaces where futures are envisioned.Components of the building systems include circulation pathways, structural matrices, lattice library spaces, and array of studio modules.

The space may be illuminated with 2D projections or 3D holograms in augmented reality. The research of the instrument is summoned in projective components that are referenced in acts of creation.

In nurturing the imagination by computation perhaps the visions are realizations of dreams at the boundary of fantasy, bridging worlds of imagination and building reality at the scale of ecology.

A model photograph of a structural framework that functions as an Instrument of Terrestrial Transformation. The framework functions as an informational matrix of environmental data content.

CHAPTER 12

Deep Storage and the Operative Architecture: Memories as Tools for Interpolative Envisioning

Deep storage may be approached literally in physical libraries or in analogies related to memory systems. This particular case study functions through the accessibility of boxes by robotic retrieval mechanisms. In following this technique, the content of the boxes is then translated into a shifting of shelves as walls that define new types of enclosures and sequence in their positioning. They double as desks and shelves and can curate content at micro and macro scales. A robotic assemblage of content includes a capacity for variant types of content to be delivered in specific mediums in a referencing of an increasing scale and access. Social protocols and operatives as capacities are derived from the measure of accessibility and malleability in a cross referencing of content. Vast storage spaces are accessed by robotic arms and platforms with their reaching expanses; the withheld content is then available through mobile wall components that store and group content in strategic ways. An expansive aggregate of rails for content to travel through is a coexistence with the memories and collections of multiple intelligent bodies instrumentalizing causal relationships of codependent data. The relationship between the content and the content itself or other users is a feedback loop of cross-referential information.

The possible types of library contents as a growing list may include:

- Books
- Biological Indexes
- Maps
- Architectural Representations
- Multimedia Displays
- Simulations
- Models
- Connectivity Models
- Biodiversity Models

- Environmental Measurement Models
- Programmatic Arrays of Buildings and Cities
- Budgets and Finance Sequences
- Lists of Agencies and Operational Structures
- Partnerships
- Educational Bodies
- Profiles
- Recordings

Below is a schematic of the robotic storage of Mansueto Library at The University of Chicago

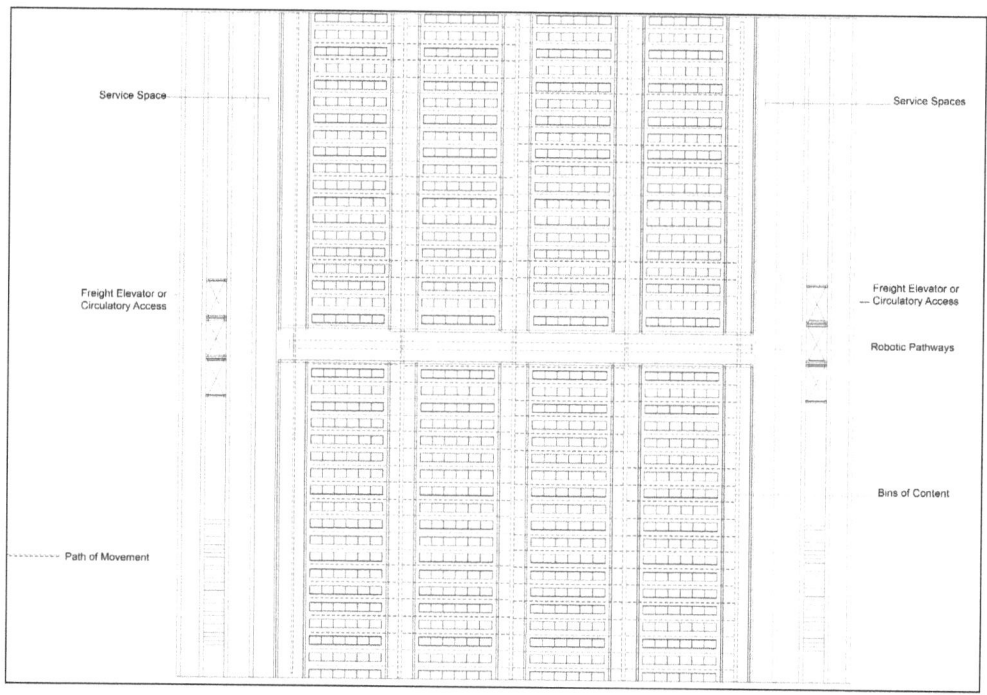

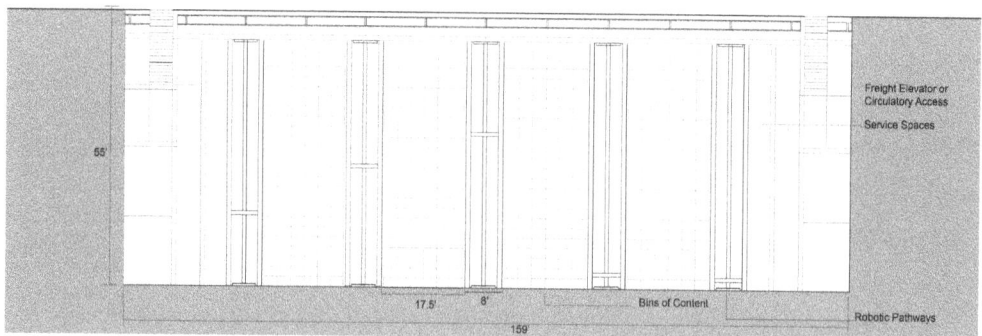

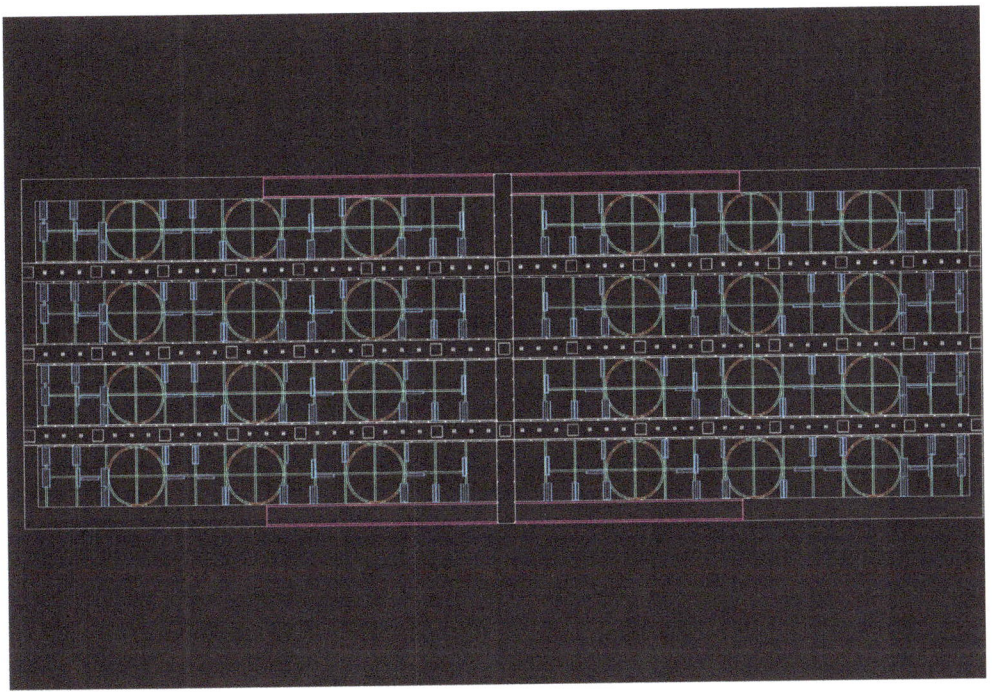

Distributive arteries (in white) deliver objects to locations where they are attributed and organized into local curatorial models of rooms framed by shelves that double as desks (in blue). Circular domains (in orange) are enclosures that are assigned to bodies of intelligence that derive a relationship with adjacent curatorial models by defining the surrounding enclosure. The screens within these domains are interfacial displays of the content above deep storage and also within.

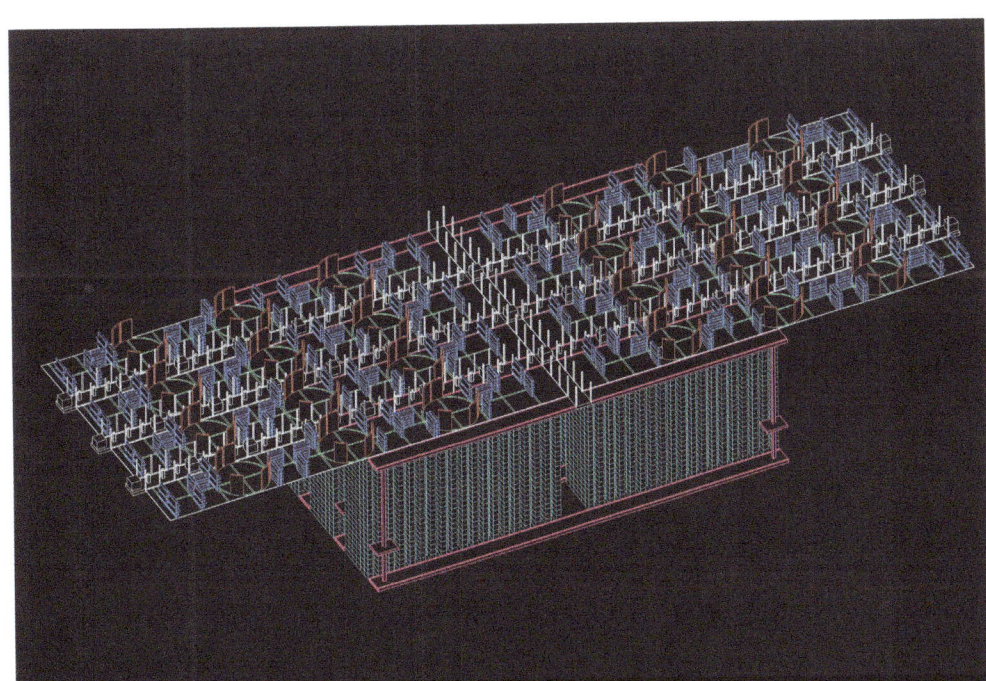

In an active role, the walls derive a type of virtuality through the construction of potentiality as an abstract space of changing partitions. The wall as a screen, withholds images. The image is a spatial, cognitive, technical world. Through simulations and drawing techniques, images can inhabit 2 to 4 dimensions. As moving walls and screens, kinetics in space are introduced into a conception of spaces and barriers. There is an additional quality of multiplicity as a means with the control of complexity in states of access. The placement of such material and cognitive substrates may delineate what Douglas R. Hofstadter calls, "isomorphic correspondences, technical identities or analogies and chain effects," in his book *Gödel, Escher, Bach: An Eternal Golden Braid.*

The correspondences in their multiplicity generate a virtual context, through a self referencing that is structured in sequence and pattern. In multiple parallels of moving walls, emptying and refilling of content defines enclosures in patterned association giving structure to initiatives and explorations. Like a canon in composition, the indexical notation of such accumulations of walls act like chords and content and notes within chords—an overall instrumentalization of the deep archive. In a canon, notes serve triple or quadruple roles being part of a melody while simultaneously being a harmonization of the same melody. Each note is contextualized by the other. Content and its correlated work is consistently being defined by its context.

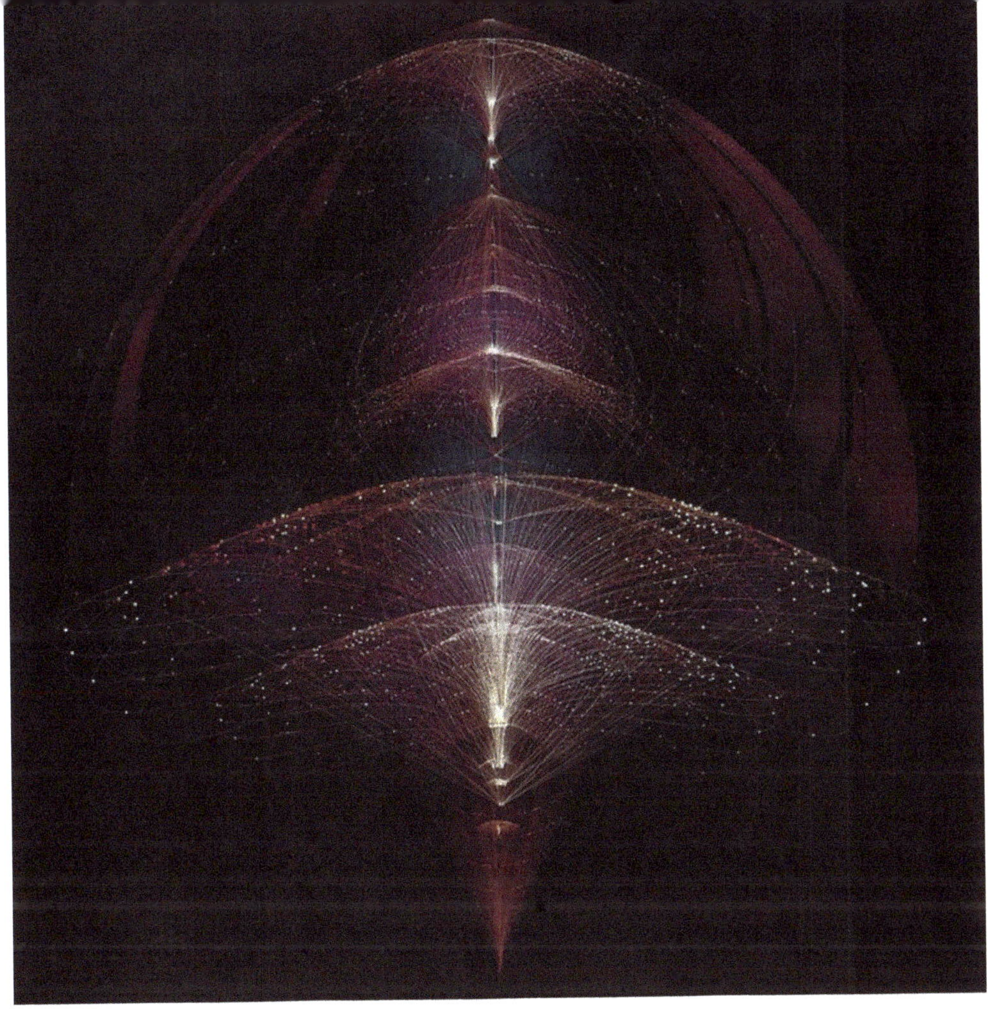

The Art of Mathematics *Tatiana Plakhova - Complexity Graphics*

Algorithmic Sequencing of Fractal Bifurcation - Array of Possibilities. Image sourced from Complexity Graphics by Tatiana Plakhova

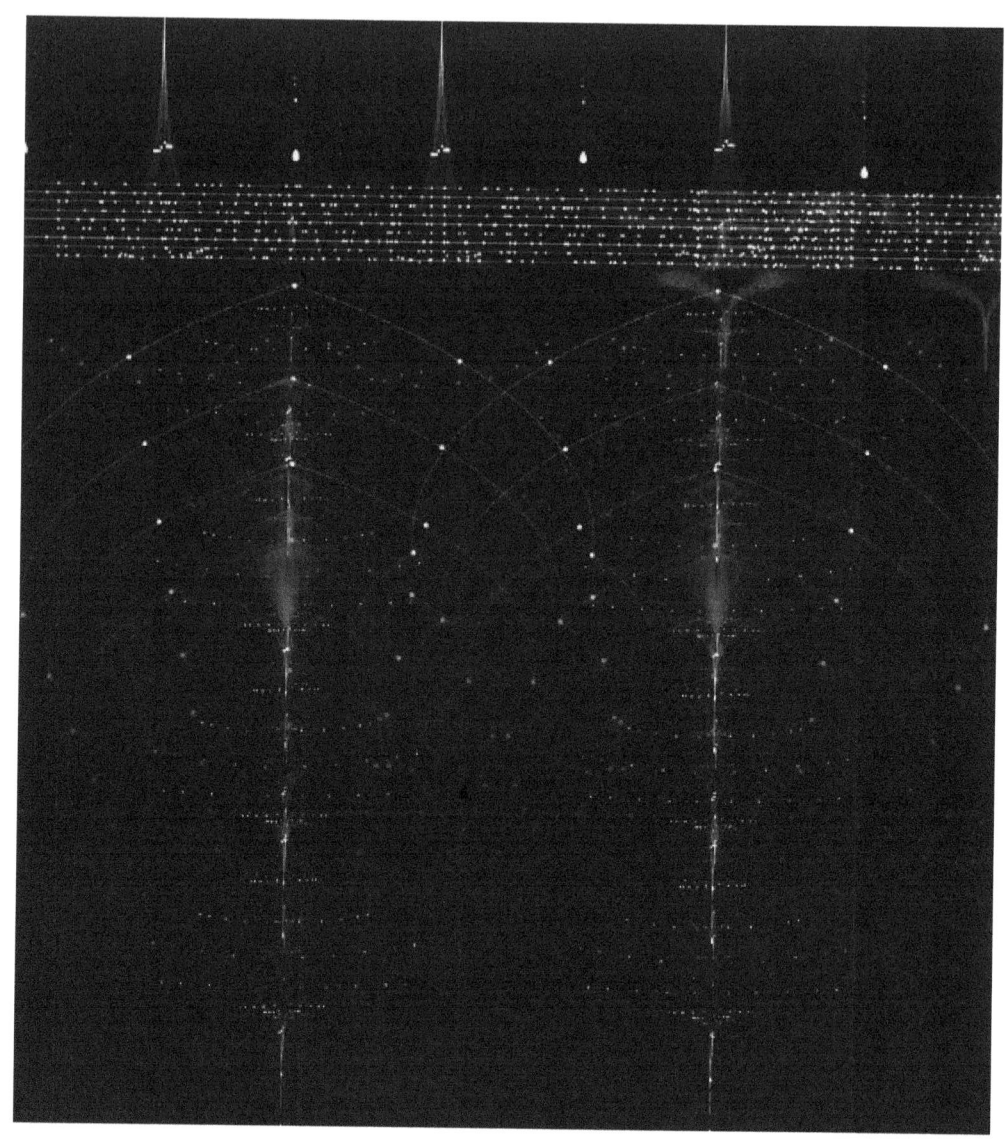

Algorithmic Sequencing of Fractal Bifurcation Translating Sheet Music.
Image sourced from Complexity Graphics - Tatiana Plakhova

What we can infer or automatically prove through the method of incremental consideration through valence or multivalence as a functional context that the instrumentalization of a single theme or component may be proliferated, supported, activated, and manipulated into a continual system with harmonic accumulations of tactical assemblies of knowledge representations. Orchestration of content or structures in looseness and rigidity afford another expression potential. Hofstadter speaks to Bach's compositional tendencies when he writes:

"His form in general was based on relations between separate sections. These relations ranged from complete identity of passages on the one hand to the return of a single principle of elaboration or a mere thematic allusion on the other. The resulting patterns were often symmetrical, but by no means necessarily so. Sometimes the relations between the various sections make up a maze of interwoven threads that only detailed analysis can unravel. Usually however, a few dominant features afford proper orientation at first sight or hearing, and while the course of study one many discover unending subtleties, one is never at a loss to grasp the unity that holds together every creation."

If the library then is to be active as a thematic structuring of themes, it is an instrumental operative potential of both the curator and the occupant (aka, student or librarian). It is a different culture of both expression and operation—both understanding and questioning—in a continually commutative re-assembly of the material content. It summons the users to obtain a certain discipline in composition towards the action.

The structural organizational capacity of the shifting enigma of the library meets a conscious meditation of the heuristic of degrees of pragmatism, dogmatism, and structuralism that are associated with curatorial models. That these models are derived from pedagogy or institutional intention are met as circumstantial objective perspectives with derivative histories of logical conclusions that can be traced and questioned.

A whole functioning of collective conscious bodies is enabled by a disciplined people that results from a disciplined environment. Trained to a mobile orchestration of kinetic screens, moving partitions, and dynamic platforms provides disciplined sovereignty over isomorphic correspondences, technical identities or analogies and chain effects as each frame in juxtaposition to the next toggles between dependencies.

As Jeffrey T. Schnapp delivered in his keynote address "Knowledge Design: Incubating New Knowledge Forums/Genres/Spaces in the Laboratory of the Digital Humanities" at the Herrenhausen Conference

147

in 2013, "The cognitive scale on which such operations have been traditionally carried out was that of scanning a small repository or collection, culling from the scanning process an even smaller subset of records or materials that then become the object of intensive analysis and interpretation, yielding a narrative, whether on the page or in space, that weaves together meanings in the lower middle (or meso-) zone between micro- and macro-perspectives, the individual artifact and the world of objects to which it belongs."

Visualizations are a craft and data are artful constructs, as Schnapp says. These crafts or interpolated projections of creative abilities with real world circumstances are correspondent constructions that are to be derived in systemic arguments of action in order to transcend from a state of memory to an operational creative act. Memories or points of conjecture and concern within a creative process have multi-valence and multiple applications. The different symbols and subject matter recontextualized surface different purposes.

The translation of memory into a library of contents, into formal systems of spatial partition and subdivision is in parallel a transformation of its text format as knowledge of content into an operational curatorial collective experience of contextual field operations. The social interaction of people occurs in the intersection of operatives by using the variant components of libraries at different times. In acts of study, production, referencing and presentation the sourcing of library material is embodied in spatial assemblies of material involving explorative action. The systemic compiling of such information or knowledge in spatial organizations function in sequence or succession, deriving a lineage of logic or a contextual sampling that gives support to a vision or study that may resemble the associative or additive structuring of a mathematical proof. The patterned assembly of content resulting in the patterning of spatial assemblies ensures cartographic repercussions of discussion and initiatives that in composition reveal an interconnected architecture of studying that may proliferate through different subjective activation.

Statements as patterns or patterns as statements that derive new vocabulary are a new cultural phenomena from time spent in the mobile deep storage library. One indirectly accumulates vocabulary in protocol of content sampling and referencing in both transmittance and reception. One's available vocabulary or available reference set of images and words is defined through a heuristic coordination and familiarity with the interface of the library as set mobile constituent displays and accessible compartments. A sensibility that the user, student, or manipulator of the library obtains functions as a telescopic lens that groups and specifies scales of context and extrapolations of individual subject matters through situations of variant depth and surface level access. The sensibility would include an operative potential of cross referencing and translation of memory states of information into fieldwork of various deployments in context with the user's knowledge of contextual relativity among other actuations in simultaneity within the referencing—providing a means of maneuvering, manipulating, and initiating chain reactive sequences of referencing within the movement of the library components.

Hofstadter uses the terminology of recursion—push, pop, and stack—and early terms of Artificial Intelligence to explain the concepts about which I am writing. Push means to suspend operations on the task at hand without forgetting where you are. Then you start a new task which is said to be on a "lower level" than the task you stopped. To pop is the opposite of push; it is closing the operation on a lower level and resuming operations on the higher level once again.

Hofstadter writes, "But how do you remember exactly where you were on each different level? The answer is, you store the relevant information in a stack. So a stack is just a table telling you such things as (1) where you are in each unfinished task (jargon: the 'return address') (2) what the relevant facts to know were at the points of interruption (jargon: the 'variable bindings). When you pop back up to resume some task, it is the stack which restores your context, so you don't feel lost. In the telephone-call example, the stack tells you who is waiting on each different level, and where you were in the conversation when it was interrupted."

Operative or creative organizational protocols are infilled with context. "For recursion is a domain where 'sameness-in-differentness' **149** plays a central role. Recursion is based on the 'same' thing happening on several different levels at once. But the events on different levels aren't exactly the same - rather, we find some invariant feature in them, despite the many ways in which they differ," writes Hoftstadter.

The same occurrence happening on multiple levels inside a geometric ordering system is a correlated recursion of repetition in variant places—the recurrence of the gallery and the library. The same experience yet slightly different ensures the contextual structure of the grouping yet the slight differences of subject matter or the capacity of activations in the manipulation of the subject matter in local and global, micro or macro potentialities—a familiarity with a large contextual system that is the stack.

Derivations are the precise following of rules from the formal systems; proofs only seem correct. Rules and axioms are circumstances of the formal systems that enable one to intuit directions and elaborations and a continual expression in validity with coherency and growth consequences. Available maneuvers for informational management are attributed with vocabularies for action that establish location within a sequence or context and a residual behavior or action with such contextual relativity. Such as:

- Axiom
- Specification
- Specification
- Specification
- Specification
- Symmetry
- Push
- Premise
- Specification
- Add
- Carry Over
- Transitivity
- Carry Over

- Transitivity
- Generalization
- Pop
- Fantasy Rule
- Generalization Specification
- Axiom
- Specification
- Add
- Transitivity
- Specification Symmetry

These terms function in the modeling of memory for prediction within a vision system and for planning within a manipulation system. The calculation of use of square footage or cycling of content in coordination with these manipulation capacities are rhythmic orchestrations, like the scanning or interval turning over of mechanical or biological systems. The rhythmic, periodic, interval recursion of shifting components affords a prediction or responsive activation of correlated pieces that drift, shift, and reactivate within curatorial activity.

An analytic protocol serves an understanding of circumstance and tangential casualties that are given by the lexicological structures of a symbolic geometrical system, giving access to manipulability. The ability to manipulate yet does not ensure guaranteed intention towards synergistic understandings. Compatibility with proportions are found and the recursive exposition of multiple layers are discovered and invented to have transgressional or transmitant performa.

Institutions may be reconstructed as a new sequence with an embedded potential to be reprogrammed through kinetic components to accommodate cross referencing of screened simulations of multiple studies directed towards the built environment in its analytics and creation project in turn. Simultaneous Location and Mapping may function in cognitive space between analytics and creativity as well as its physical responding to such referencing of any subject matter that remains in a dependent flux, an integral quality of the responsibilities that a research and creative laboratory may be responsible for.

Possible arrangements becoming tangible separations giving rise to a virtuality through the curation of information. The virtuality affords chronological constructs of events and theories and new parallels to their coexistence especially under the new light of interdisciplinary conversation.

Morphology: Potentials of Prototypical Abstract Geometry

Geometry is posited with capacities and abilities, serving many purposes for the architect and those concerned with the environment. The following chapters in Part 3 explore the capacities of geometry as in their abstract form, how geometry is applied to real world conditions. The former parts of the book that speak about data infrastructures and biospheric systems preface the text about geometry as geometry is a prescribed solution for the context addressed by the biosphere and data infrastructures. Geometry is capable of bringing order to complex compositions. As data must be aggregated and the energy of the biosphere must assume a shape, geometry plays a role in the formation of such realities. With such a crucial role in the tool kit of the architect, geometry can be used to articulate the nature of morphological constructs in their essence and all of the vectors of evolution. The following chapters seek to describe the key points of investigation within my morphological considerations. More than the final resultant form having meaning, the process by which a form is generated is discussed as well its potential for transformation. The intersection of urban form, architecture, and landscape architecture is examined, highlighting the role of architectural languages in mediating space. The drawings are described as languages and grammars that represent specific relationships between landscape and architecture, capturing ideological descriptions of architectural constructions, along with an emphasis on the societal consequences of architectural delineations.

Part 3 of this book explains how pathways, floor plans, cartographies, and models articulate social relationships within the built environment. It delves into the conceptualization of abstract geometry as a means to emulate mechanical and biological systems through morphological principles. The role of grids and geometric organization in both natural and artificial systems is discussed, emphasizing the importance of computational operations such as

155

addition, subtraction, multiplication, and division. Part 3 introduces the concept of "Harmonic Grids," which explore complexity and robust interdimensional relationships. These grids create parallels, alignments, intersections, and divergences, defining enclosures and volumes. The interplay of multiple layers and organizational potentials is crucial in understanding how systems can integrate and interact. A focus on organic systems, where the geometry becomes less conventional but still rooted in fractal-based cellular structures, is throughout this section. Parallels are drawn between organic structures and solar radiation, emphasizing the morphological and chemical relationships that define life forms and their environment.

CHAPTER 13

Geometric Grids, Systems + Organic Morphologies and Crystallography, Analog and Digital Drawing and Modeling Techniques Following the Rise of Digital Architecture

began drawing in my third year of college at Southern California Institute of Architecture to explore the potentialities of forms and formal languages as I was consistently surrounded by renders, models, and architectures that expressed different variations of formal vocabularies. The drawings were an excellent vehicle to explore geometry and morphology in an experimental way without necessarily adhering to a program or studio brief, allowing me to focus on more personal concerns regarding art and architecture. I was reading philosophy, architectural theory, and history and doing well in my design studios, and in my spare time exploring the potentialities of my drawing skills, which were not being taught in class. At certain developments in my studies the drawings started to become applicable to my studio briefs. The different design studios I was enrolled in demanded versatile ways of thinking and technical expertise to achieve solutions for diverse problem sets. I was exposed to forms of computational architecture—discrete cellular automata, parametric semiologies, and ecological thinking—while also reading about different environmental, biological, mechanical, and object-oriented philosophies. I considered the experience as an exposure to a vast constellation of knowledge, an epistemological universe. The drawings helped me to navigate the educational realm with an integral continuum, with certain fundamental principle answers persisting as the questions changed. The fundamental truths of my architectural thought began to surface and useful methods proved themselves relevant in the problem sets I was facing. In retrospect, my drawings have the capacity to be categorized to technicalities of formation that describe the operability of systems as rule sets and algorithms that give rise to certain structures and forms. The following are categories of geometrical operations that I was exploring:

1. Tetrahedral, Orthogonal, or Radial - Grid or Field Systems
2. Second and Third Dimensions
3. Organic Systems, such as membranes, structures, tissues, surfaces, topologies, and fibers

Geometrical Grids and Organic Morphologies

During my thesis presentation at the Harvard Graduate School of Design, architecture theorist and professor Michael Heys used the terms "prototypical abstract geometry," and I found this terminology relevant to discuss the drawings in terms of their potentiality. At another talk at Harvard, Henry Cobb, Rafael Moneo, and Peter Eissenman discussed abstraction as the basis from which a future architecture would be generated.

The drawings are inherently procedural, following operational rules to generate a result. Eissenman's procedural drawings may be considered precursors to parametricism or discrete cellular automata as his processes were indexicaly recorded to generate his architecture, the procedure may be considered to have a certain code or language to describe the operations. His architecture was generated from ideal concepts of geometry and proportional grids which differentiates the procedure from a site driven design like Bjarke Ingels. Eissenman's diagrams are geometrical operations that can be described using the language of mathematics, multiplication, division, subdivision, areas, angles, distances, vectors, direction, midpoints, centroids, and reflected such operations in particular forms of drawing, planimetric, or axonometric projections while Zaha would use perspectival drawing techniques as projections of space inside the two-dimensional paper realm. Both Eissenman and Zaha allow their forms to drive transformation in an ideal field space; there is an internality of the form and its qualities that drive the formal operations. Similarly, Sol Le Witt's incomplete open cubes are not necessarily a procedural construction, but notate the component based formation of a total whole.

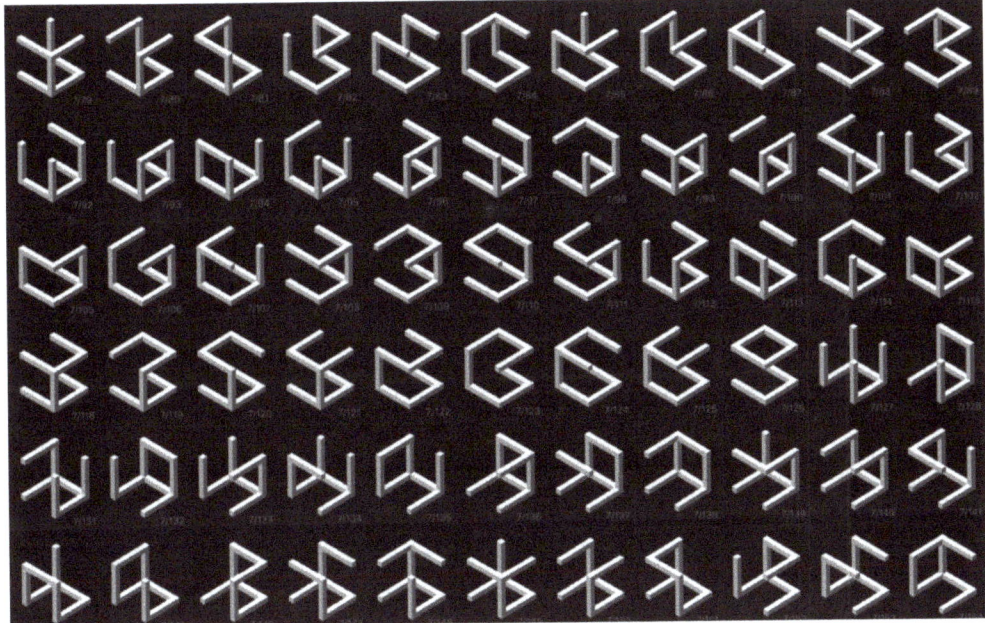

Image courtesy of Rob Weychert

Paintings or drawings by Zaha Hadid also explore an intersection of intuition and rule based procedures. It seems her early paintings followed intuition, while the ability to calculate the geometries began to post rationalize the morphologies into parametric systems. Her paintings often consisted of vectors and trajectories of movement that were generated from uses of the building as circulation or enclosure, following the ideal geometric paintings the dynamics in the site represented in linework and surfaces to articulate a form that would respond to intersecting uses.

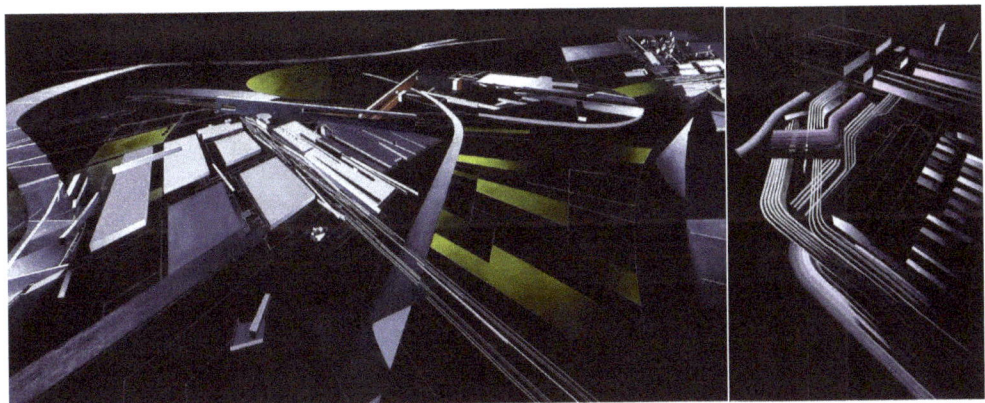

Zaha Hadid Paintings © Zaha Hadid Foundation

Abstract geometry serves a certain potential in its innate qualities that imply future elaboration and construction; a building massing is an abstract simplification of a building that can imply the location of the building, the definition of its interior exterior and facade. Figure ground drawings of cities are abstract to discern negative and positive spaces. Abstraction is so powerful that a simple grid projected onto Long Island can infer all infrastructure and building zones and locations.

At the intersection of urban form, architecture, and landscape architecture could be derived a projective ability of architectural languages to mediate space in a synergetic medium. Space is the product of what we perceive and what is presented. Perceptions of the intersection between architecture and the larger scale acts of landscape and urban territories are mediated by languages of representation that depict specific relationships between landscape and architecture, architecture as an urban participant. Drawings use languages and grammars through their structure to discern the behavioral relationship of objects that are drawn, creating a situation of subjectification that can be compared to simulation. Early two-dimensional drawings of certain architectural projects can be used to describe a particular subjectivity of landscape to architecture. Ideological descriptions of architectural constructions are captured through their representations that directly address constituent elements within the drawing.

In a two-dimensional understanding landscape and architecture can be described in a manner of cohabitation; yet, what is simultaneously drawn is a participatory societal role of the individuals who occupy both the architecture and the landscape. The continuation and proceeding precession of pathways, symbologies, and floor plans articulate behaviors and actions in which people participate. The architectural description of pathway and partition is fundamentally instrumentalized in a strategic action to articulate social relationships between individuals. The landscape in its context to be transformed into architecture is then understood as a malleable surface of manipulation by the acts of drawing that delineate transformations of both the landscape and the architecture simultaneously—acts of delineation having societal

162

consequences simultaneously.

The canvas or landscape that is augmented according to *"events and circumstances"* is traced through a simultaneous phenomena of representation. What David Harvey describes in a "A 'Grid' of Spatial Practices" are material practices, "flows of goods, money, people, labor power information; transportation and communication systems— accessibility and distanciation, appropriation and uses of space, domination and control of space"

With such events and interchanges of energy distributions of power and influence that augments landscapes and urban facility there is a simultaneous representation of such augmentations in representations of space as Harvey describes. Cartographies, media and drawings delineate the course of such events and their ecological influence as the urban and the landscape are transformed. Through careful extrapolations of elements and their use in architectural and landscape architectural descriptions, strategic methods of representation may be created to understand the methods in which representational delineations shape the physical boundaries of architecture and landscape and the sociological phenomena that occupy such delineations.

The generations of architectures that are embedded into the landscape and cooperate together through simultaneous strategies are generated through the instrumentation of representational substrata that delineate the possibility of cooperation between the built structure and the medium of representation that facilitates construction, space is what we perceive. The constructions themselves exist between a real and imagined space created by the imaginative ability to draw the landscape and the architecture simultaneously. The spaces that are designed by architects, landscape architects, and urban designers exist as spaces where imagination and memory are combined.

As an example, let's consider Hadrian's Villa Adriana, which has distinct elements of architecture and identifies the overall model of the villa as a potential urban condition. The elements are constituents, drawn to each other's meeting through an articulation of linework that frames the programmatic performances and captures a cohabitation of "city like" dwelling spaces. Information is illustrated to show a demonstrable context of the fitting of such discretized pieces into larger constituent units of planned architecture.

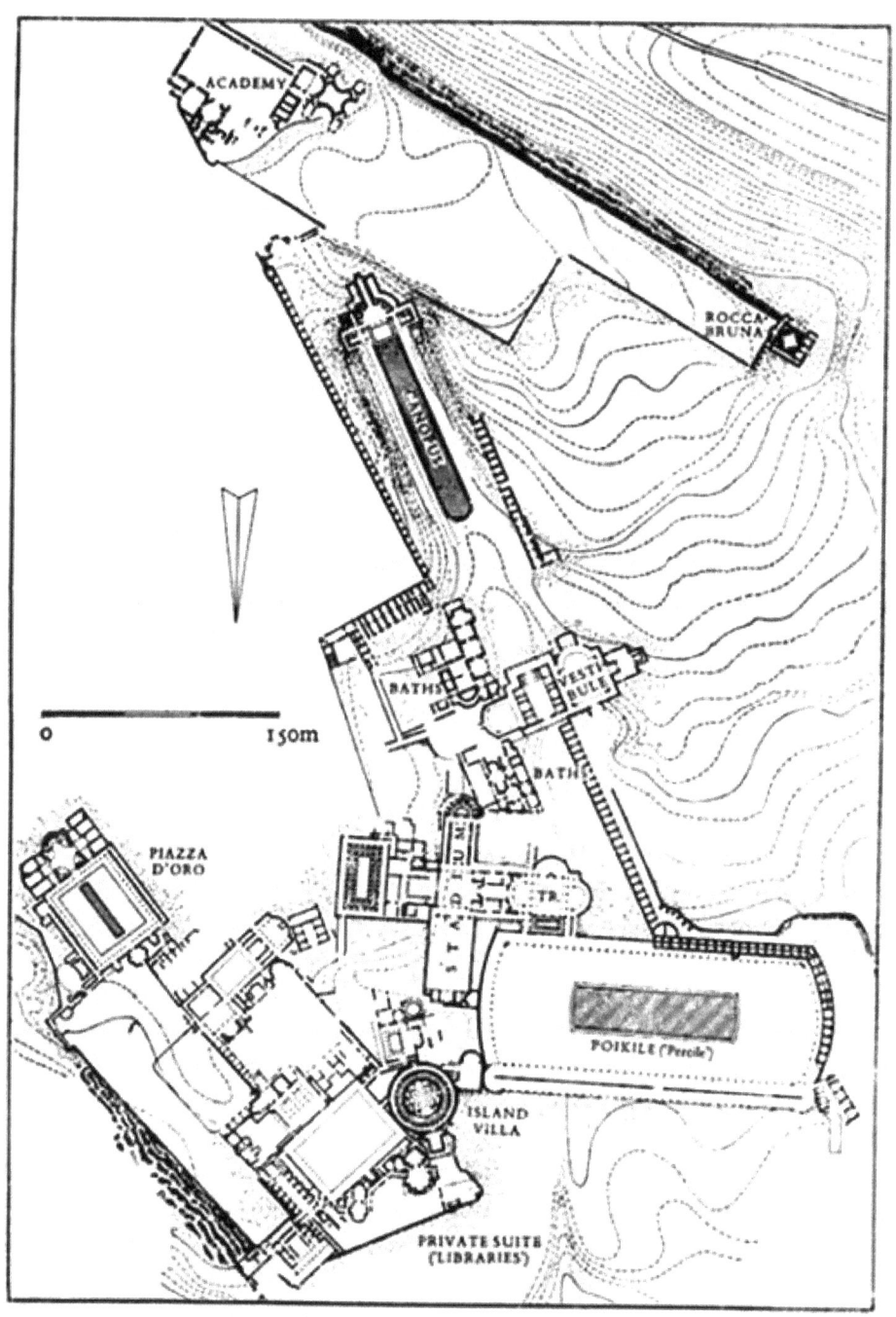

Plan of Hadrian's Villa. In James Ackerman, The Villa: Form and Ideology of Country Houses. *Princeton, NJ: Princeton University Press, 1990. 51.*

Abstraction, as a geometrical system describing urban and architecture scenarios, was realized alongside the concept that one can emulate mechanical and biological systems with morphological principles. The functionality of the environment being both biological and mechanical seems to beckon generating architecture through the fundamental principles within these systems. A principle of both the biological and the mechanical in the urban or architectural context is a gridded organizational scheme, which through its delineation organizes behaviors and functions. Imagine the grid of a leaf or the motherboard of a computer to arrive at the concept that conduits of information and energy use a morphological matrix. Computation was formulated on concepts of structure and sequence according to definite rules and units. Computing and calculating involves the geometrical operations of addition, subtraction, multiplication, division, subdivision, areas, angles, distances, vectors, direction, midpoints, centroids.

Tetrahedral, Orthogonal, or Radial—Grid or Field Systems in Both Second and Third Dimensions

A geometric grid or field was one of my first explorations in architecture school where I discovered the construction geometry behind the Hagia Sophia in radial proportions: the poche for the arches, columns, and domes aligned to a geometrical grid. It was clear to me at an early stage that architecture is easily generated from such planimetric organizations. In harmonic grid studies, my concern with complexity and robust interdimensionally was explored to arrive at both discipline and freedom, simultaneously expressed in space. The geometries create parallels and alignments across distances, intersections, and divergences while also defining enclosures, volumes, and figures that are adjacent, contingent, interconnected, and interdependent. The harmonic grids are formed by inscribing tetrahedral geometries in tangential circles that are arrayed in the x and y axis. The tetrahedrons align to each other and are sometimes also perpendicular to each other causing moments of orthogonality. The multiple layers and potentials of organization I consider useful at

165

macro organizational levels when considering how multiple systems can become contingent and integrated. Informational matrices of continually increasing complexity while maintaining a disciplined structure are composed through studies of algorithmic proportions. Grids are projected in spatial ratios, intervals that are proportional, generating recursive relationships, otherwise described as self referential. The components of the grids are stratifications that stretch vast distances while being subdivided in relative alignments. An array or constellation of points with inherent organizational principles are projected into structures of organization forming an interdimensional space of expanding axes of mobility or freedom. Relational frameworks of lines are projected spatially within themselves. Perspectives are nested within perspectives, volumes within volumes with proliferating organization across layers. The genesis of the form is imbued and encoded with organizational principles that discern its future assemblage as it is animated.

Geometrical components as points, lines, and surfaces are pivotal moments of change within a system that are available to our degree of control and augmentation. As the geometrical components amass, their distinct locations reveal a functioning of the system thereby affording an array of usage possibilities.

A geometrical system consisting of many systems within itself implies its potentiality as the combinatorial relationship of components defines a functional formation as the systems of points, volumes, enclosures, lines, spacings, and axes infer relationship to each other. A geometrical system can be generated by and function as a mechanism. Or as Siegfried Giedion writes in *Mechanization Takes Command*, "A mechanism therefore consists of movable parts that work together and periodically return to a set relation with respect to each other. It consists of interlinking parts, each of which has a terminable degree of freedom of movement: for example, both a pendulum and a cam valve have one degree of freedom of movement, whereas a threaded screw has two. The fact that these varying degrees of freedom of movement can be quantified means that they can serve as tangible guides for measuring, for setting limits on the amount of

movement that can be expected between any two interacting solid objects.

In every machine, then, movement is a function first, of the way the parts interact and, second, of the mechanical operations of the overall unit."

In my abstract drawings the relational movement of components is imagined as a dynamic formulation of space. The drawings are derived and formulated and thereby constructed through a dynamic process of emulation as a procedure in order that certain systems of components like an array of points and lines may be proliferated to be in existence with relationship to one another, that their existence is identical to their contingency. However the dynamic formulation of the drawing is captured, finished, and stilled in a seized moment and location in time so that all moments of time that the drawing was formulated in are able to be seen a single moment, as if one saw a song instantaneously in its completion in one moment and understood its harmony. As the drawing is formulated it moves through variant phases and states that are driven by its rule sets; the state of the drawing carries a certain measurable value as relationships are defined to inform the decision making of following steps in the procedure.

A fundamental example of relational components inside a mechanism is the lock of a door. The inner functions of locks used pin tumblers, thin rods that move in locations inside the lock mechanism depending on where the key determines their location as it slides the tumblers along their path displacing their fixed locations as toggles that engage the transitional states of plugs that retain closed and open states.The fixity of a material tumbler or pin gains a merit of abstraction as in use of fixed points in space or the mere sequence of components of a mobile system with which the system may achieve a stasis in static arrangements.

Giedion writes, "To unlock this mechanism, a small flat key is inserted into a narrow slit in the plug. It passes under the downward-pressing pin tumblers and raises them to a point at which the junctures between the pin tumblers—divided as they are into two separate functions—correspond exactly with the juncture between the lock case and the plug, that is just between the fixed and mobile cylinders. When they have reached this precise point they no longer oppose the rotation of the plug, which will now go around—as one turns the key mechanism is unlocked."

This description of the geometrical component performance is entirely materially dependent on an arrangement of physical components. The platonic language of mobile cylinders and pin tumblers initiates the conversation in the restrictions of a material world and a toggle terminology is introduced in this circumstance as limited to its material substrate. Geometrical component stasis within a dynamic formulation of a system is inseparable from the concept that our multidimensional reality is composed in states of being. The interior of a state includes the ability of that state as an available set of relations inside the stasis as in the morphological ability of water to have certain surface tension when it is a liquid so that a water strider bug can walk along its membrane, or the state of a line to be straight so that a perfect perpendicular line may be projected through it, or in more complex terms the state of a frame enclosure being completely enclosed in order to derive that its edges can be projected from its locations to meet other enclosures and form a link of enclosures as in the parcel system of our cities. In abstract terms, the state of a geometrical figure at some point of its formulation process contains properties that afford, and even logically infers, its transformation, especially in the case of a cybernetic mechanistic organism hybrid.

Radial grids are also concerned with organizational ability; however, these grids address the dynamics of moving components in circular and linear fashions as an architecture that would assemble and disassemble in a more literal fashion than abstract grid formulating in geometric relationship to itself. Walls configure as they shift and move along the delineated tracks in the drawing. The addition of radial geometry defines zones within the grid that are circumscribed. Sequential centers and nested centers are another form of relationship. Axes of movement are delineated with straight lines having degrees of preference in the rest of the system due to their width as an artery in the body of an organism or a main avenue through the city, adorning components of pathways. Enclosures intersecting and branching are stemming and relating to a

168　fundamental armature of the structure as a skeletal system hosts muscles and nervous system structures. Just as the keyhole holds a host function for the tumblers and the constituent components of the lock. Other examples include conduits and movements of information through space in circuits as in the computer or pipes as in cars. A drawing that uses both enclosures and strata of conduit movements simultaneously can form a field of systematically composing flows of movement and states of enclosure. In the abstract formulation of geometry, local and global conditions are relevant to emulate specific constituencies of individual components.

By using 3D modeling software, we can rotate the view of the grid so that the z axis is made visible to open the tetragonal, or trapezoidal, orthogonal structures to span in space and achieve new adjacencies. A subdivided grid projected into a perspective space can be manipulated in reference to geometries that are projected through it and off of it into adjacent space. Points of reference imply an existing dimensionality of space, informing the construction of new volumes that can be placed in three dimensional space while aligning to each other and completing other vectors and directions by adhering to the original matrix and its deviation patterns. The volumes that expand in reference to a grid can then achieve certain qualities and properties of their own defined by the angles of their enclosed surfaces. Once such independent qualities are established, new dimensionalities, surface normals, vector directions, and potentialities become available in response to the newly established volumes inscribed in the perspective grid. With multiple vanishing points, the implied vectorial directions of space can be proliferated and the subdivision of surfaces that follow such vectors brings measurable dimensionality to an architectural space. Multi-valent, interdimensional adjacencies are axes for the geometrical system to be expanded upon. Multiple operations function simultaneously as contingent geometrical systems interact with each other forming alignments across space and surfaces and subdividing large volumes, defining local and overall structures that are in communication with themselves. Analytic geometries allow one to subdivide surfaces and measure dimensionality or reveal surface normals and directions in order that operative manipulations of the geometry may take place. A continuous process

169

of defining geometry as it expands and elaborates and a self-referential analysis to define properties and qualities of the geometry that inform its elaboration takes place.

In other circumstances the drawings follow the organic interpretations of membranes, structures, tissues, surfaces, topologies, and fibers forming more anatomical conceptions of form; however, the geometric principles of analytics are still relevant in the sense that a structure and surface may be derived from similar principles and in construction delineate codependently.

The scaleless quality of fractal-based fibrous geometries reinterpret the differentiation between an organism and the environment as blood vessels resemble river systems and tree roots across the forest floor resemble nervous system architectures.

The definition of a membrane happens in a certain tectonic subdivision of a surface as in the tension of a bubble or the facets of a crystal. The lines that define a surface can also be projected into space to become structured in the same way drawing and architecture and multiple fibrous lines can accumulate to form both surface enclosure and volumetric geometry.

The definition of components and their interconnections is fused in the total system, an interdependency in morphology as in a manifold layering of the nervous system. There is recursivity as fibers and members of the geometry expand in reference to themselves, cycling bifurcations and deviations, expansions and connections with other members that are all integral. The different organic systems within the body are in communication with each other in chemical interactions that have morphological consequences. For example, joints are a combination of bone structure, muscle, and synaptic nervous system fibers that allow bending and manipulation of the body.

In *Cybernetics* Norbery Weiner writes, "The central nervous system no longer appears as a self-contained organ, receiving inputs from senses and discharging into the muscles. On the contrary, some of its most characteristic activities are explicable only as a circular process, emerging from the nervous system into the muscles, and re-entering proprioceptors or organs of the special senses. This seemed to us to mark a new step in the study of that part of neurophysiology which concerns not solely elementary processes of nerves and synapses but the performance of the nervous system as an integrated whole."

In organic systems the typical geometries of conventional mathematics are less traceable. It is more difficult to discern primitive figures like circles, squares, and lines that most conventionally form grids in a distinct pattern; yet fundamentally, there is a geometrically fractal-based cellular nature of all organic material. Linked to solar radiation the cells and structures of organic material are embedded with the capacity to transmit energy and leverage the capacities of morphologies that allow the energy to be consumed, expressed, and transformed. This residual relationship of morphology to chemical energy is fundamental to the expressive forms of life as plants and animals that accumulate to create an environment all have a relationship with their surroundings. The different functions of biology are also the different potentials of prototypical abstract geometry. As discussed in previous chapters, the processing of information through morphology is one potential.

Epistemological information may be spatialized to organize relationships and associations of cognitions. The fundamental inner workings of intelligence and computation are interoperable systems of logic that can be expressed as geometric systems. These interoperable geometrical systems can proliferate and grow to be vastly complex as systems stack and nest between each other and become dynamic as the relationships transform. Both geometrical systems and music have this dependent correlation of components within a system that align to and communicate with each other either in terms of defining each other's future transformation—as in the course of a melody or the trajectory or arc of a curve. Johann Sebastian Bach used algorithmic principles to construct his melodies in order to derive a structure of notes that would formulate in dependency themselves. Composing multiple algorithms to cofunction in layers is a shared principle both in computation and music.

As there is a certain procedure to arrive at the large amassing of unified composition, the procedure as indexical recording of operative manipulations of the subject composition as shown in Eissenman's drawings, as an index of procedures, may be compared to the development of mathematical proofs. A mathematical proof even in some of its most simple forms as the proof of an algebraic equation is a derivation from incremental relationships of numbers. An algorithm of geometry is rooted in the concept of the derivation as a step-by-step procedure indexically recorded to explain relativity in its elaboration. The mathematical proof and indexical procedure are strung together to form steps of reasoning in the transformation of formal systems.

The documentation of an architectural geometrical process can be indexically recorded, but the total intelligence of decision making may not be entirely present in the indexical diagrammatic process. Through the accessing of intelligent mechanisms in organisms as natural intelligence evolved over billions of years and layered systematic transformations occurred, the organisms acquired latent and abundant levels of interconnectivity. The intelligence of nature is observed as the interoperability of the functioning mechanisms that exist in all organisms that accumulate in the environment.

Crystallographic systems and organic systems in terms of algorithmic formulation share principles of formulation. A fundamental property of the crystallization morphological process is the relationship of planes in space as relative inclination angles. Edges are defined by the meeting of planes while the direction in space of the surface normal and the edge are relevant to the composition of the whole system. A sphere of projection creates a spatial outline, like a grid in which axes of symmetry may be used for derivation. Within the spherical projection system there are also interiorities that are proportional to the overall sphere. There may be projection spheres within projection spheres as concentric spheres. In-between axes of symmetry are zones that define areas for morphological interaction in particular reference to other parts of the spherical projection. Zones may be adjacent, opposite parallel, perpendicular, or have some form of communicative angle. The axes of symmetry also define planar reflections inside the spherical projection system. Within this spatial matrix of axes, nodes, intersections, and alignments are origins of

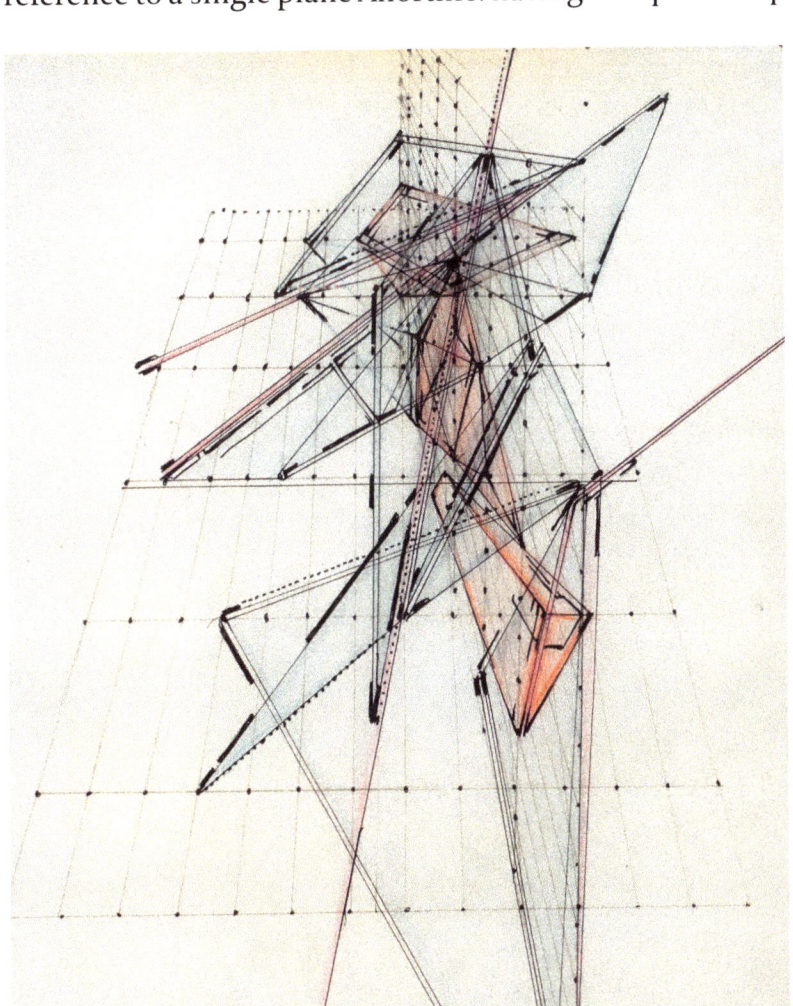

forces that affect the manipulation and connection of paths of formation. There are degrees of symmetry depending on how many axes are reflected upon. Here are some definitions of the most common types:

Cubic: of or denoting a crystal system or three-dimensional geometric arrangement having three equal axes at right angles.

Tetragonal: of or denoting a crystal system or three-dimensional geometric arrangement having three axes at right angles, two of them equal.

Hexagonal: designating or pertaining to a crystal system in which three coplanar axes of equal length are separated by 60° and a fourth axis of a different length is at right angles to these.

Orthorhombic: of or denoting a crystal system or three-dimensional geometric arrangement having three unequal axes at right angles.

Monosymmetric or zygomorphic - symmetrical bilaterally with reference to a single plane Anorthic: having unequal oblique axes.

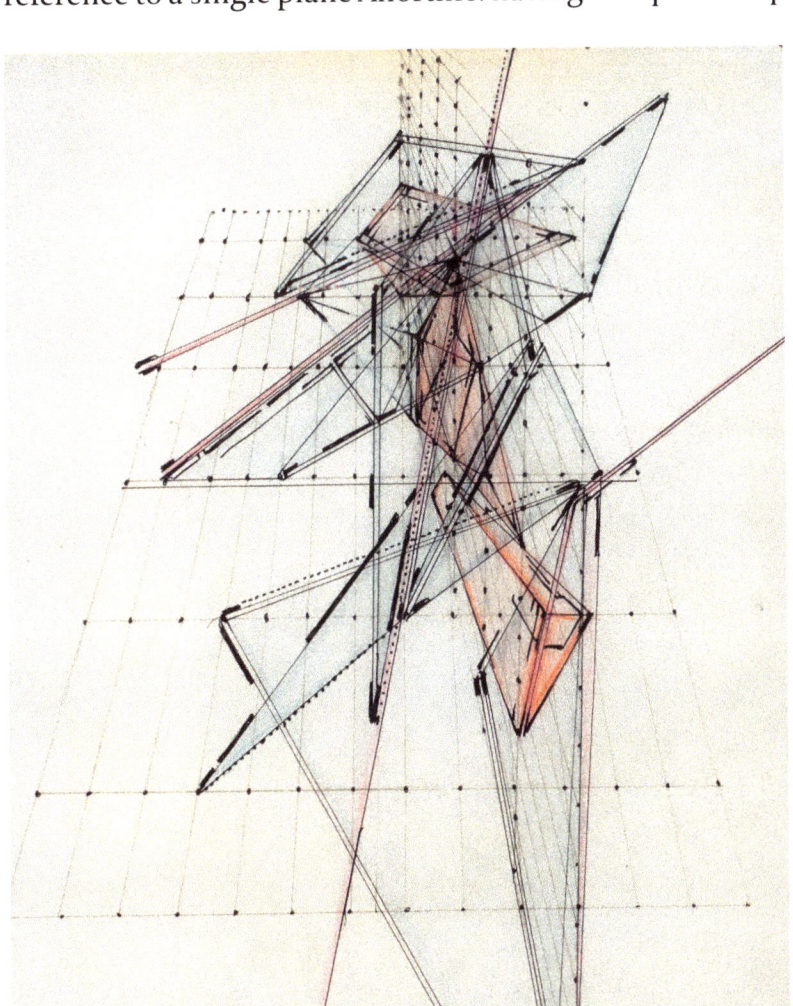

From my sketchbook, a string of tetrahedra structures projected along a surface grid.

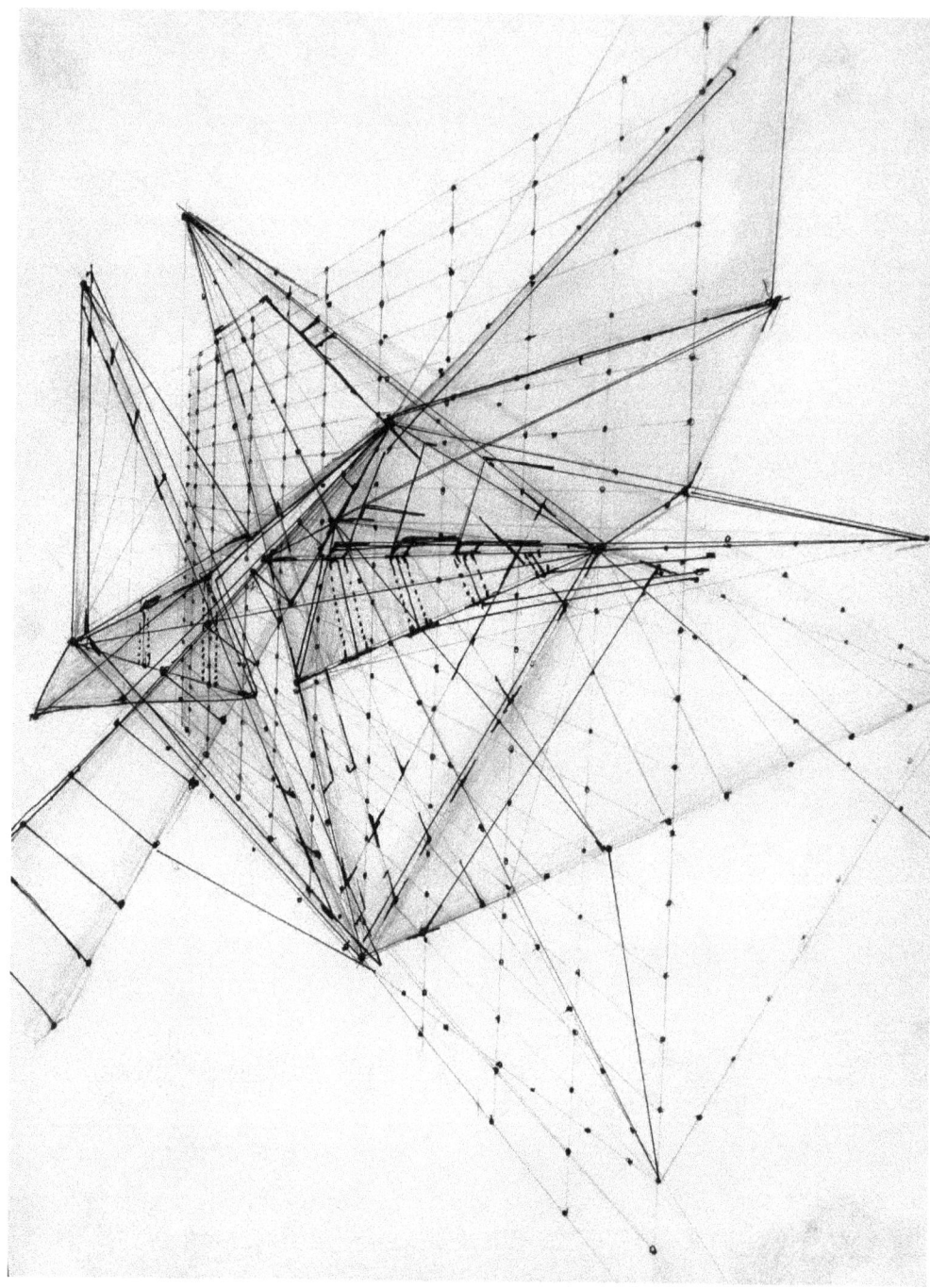

From my sketchbook, a string of tetrahedra structures projected along a surface grid.

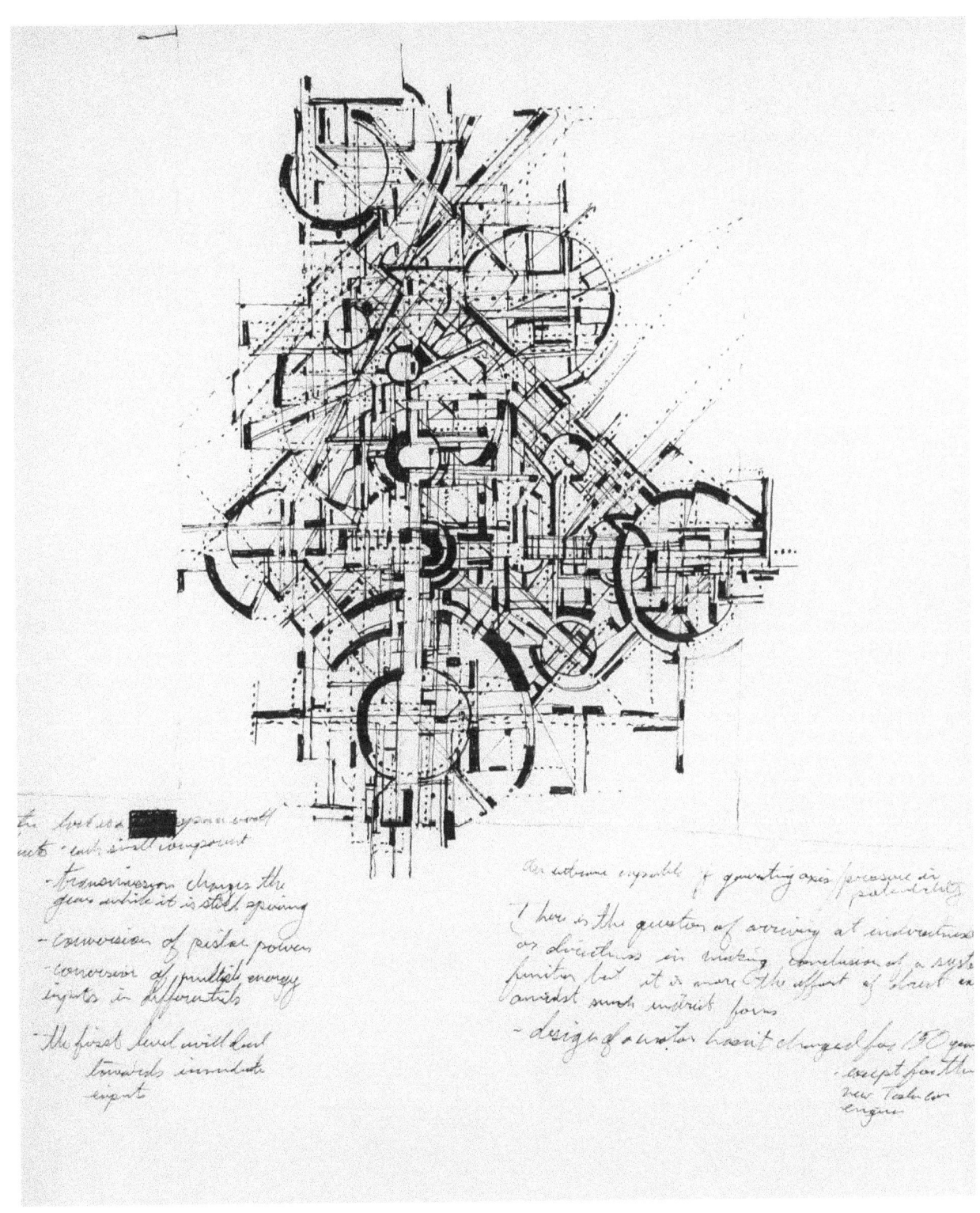

From my sketchbook an abstract diagram for a robotic building with shifting

components like the gears of a clock or the components of a lock.

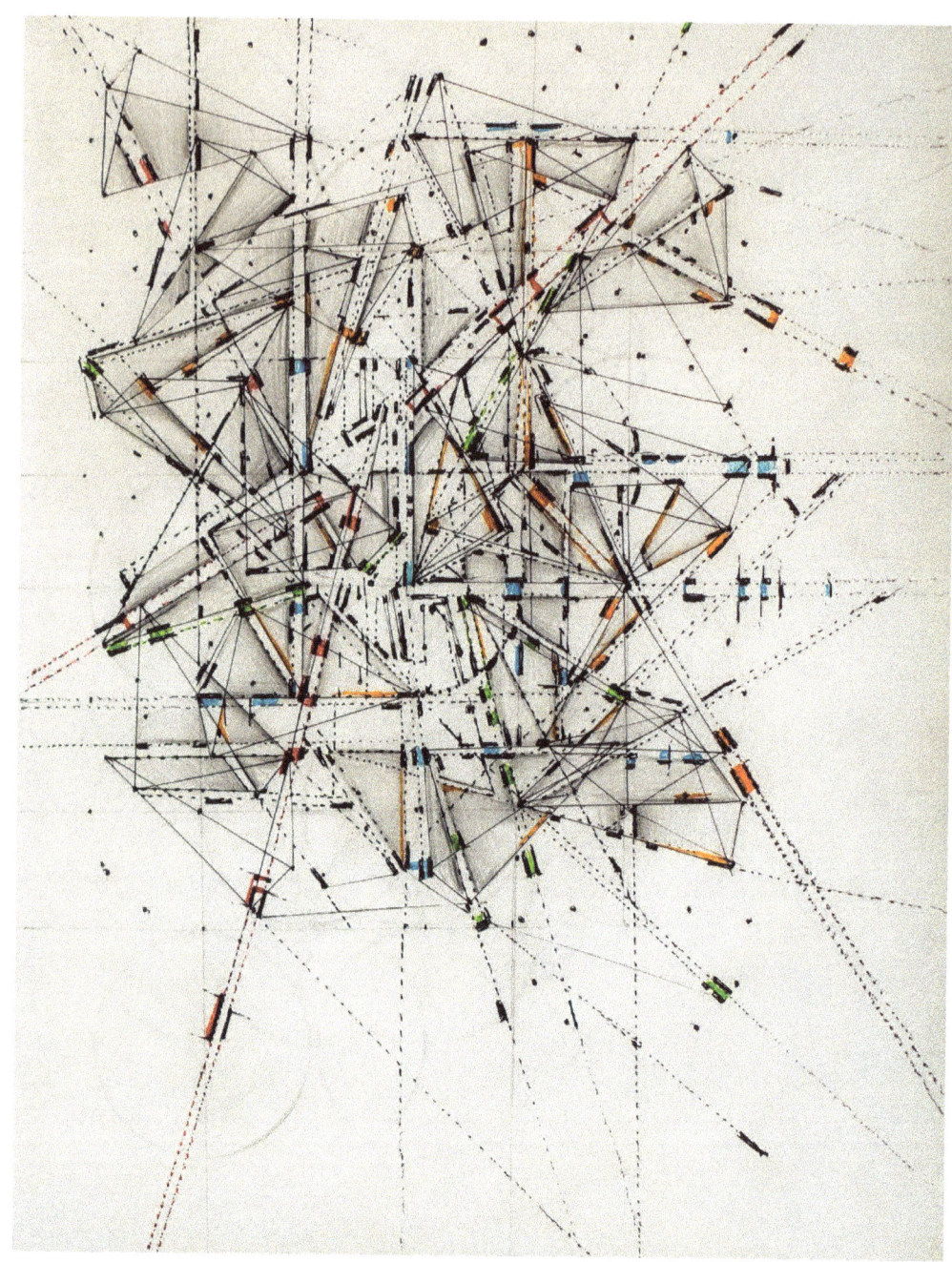

Tetrahedral grid with many converging axes, seeking to display the superimposition of chord structures in music.

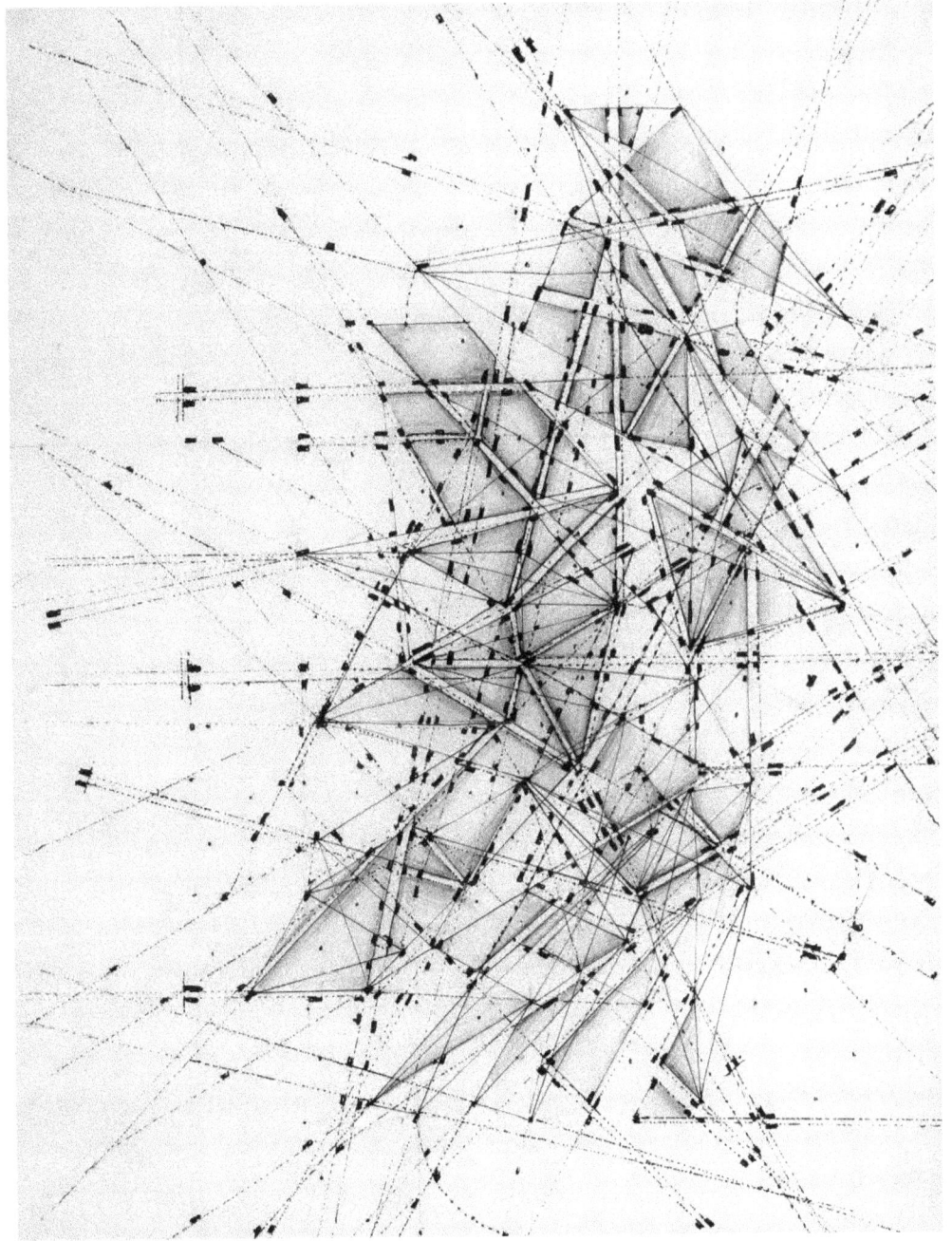

Tetrahedral grid with many converging axes, seeking to display the superimposition of chord structures in music.

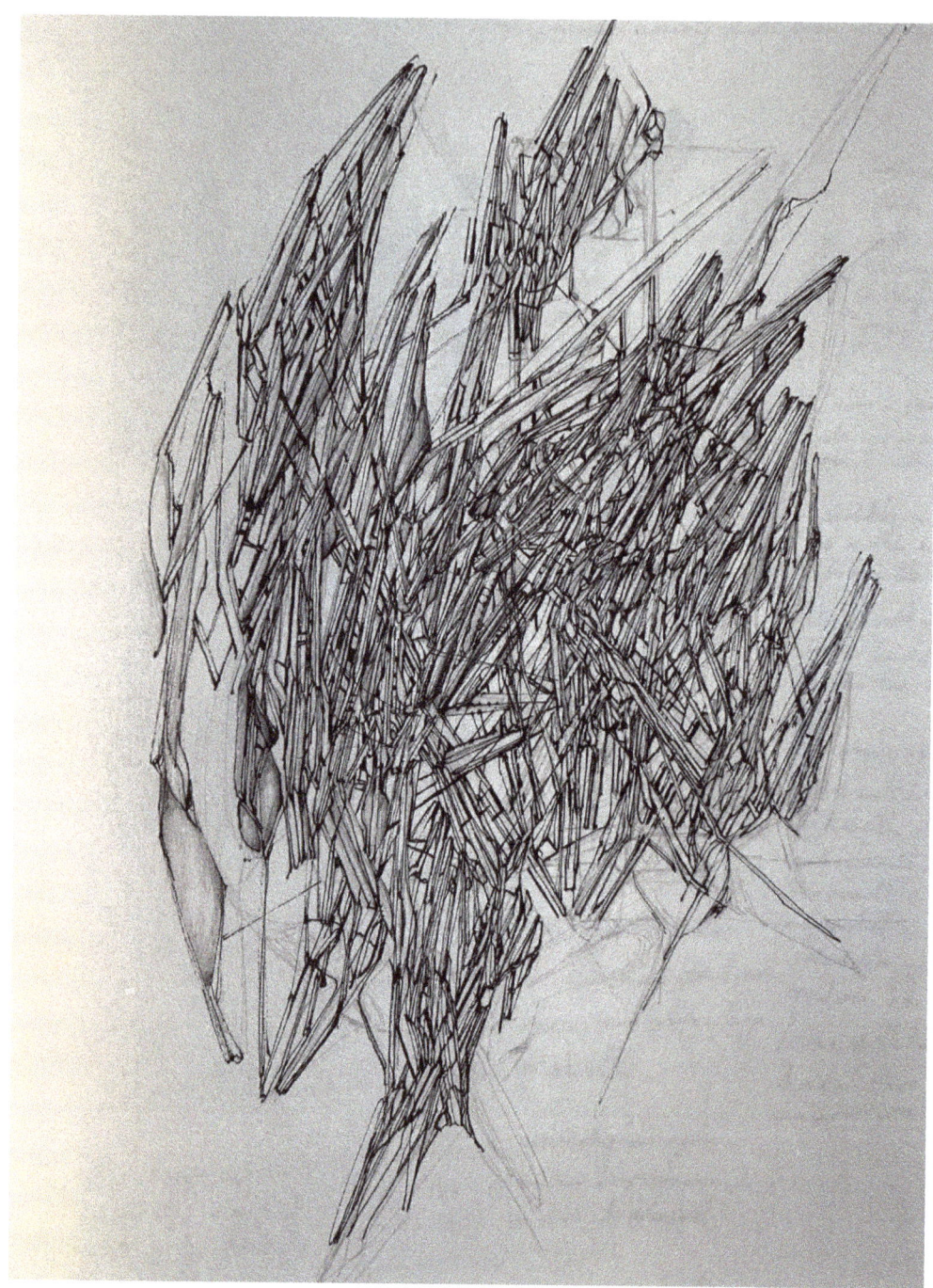

Abstract representation of organic tissues and membranes

Looking up at a crystalline tectonic structure embedded within concentric rings

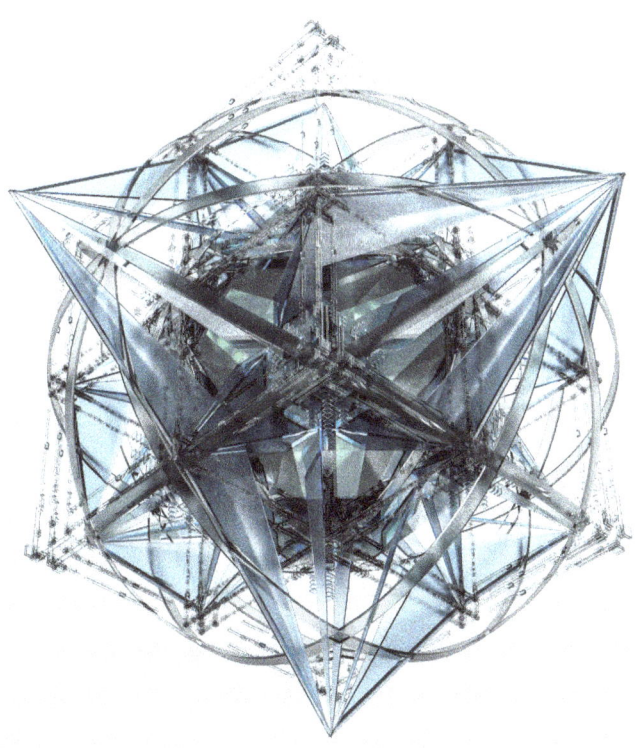

An exploration into the geometry of a cubic crystallographic system

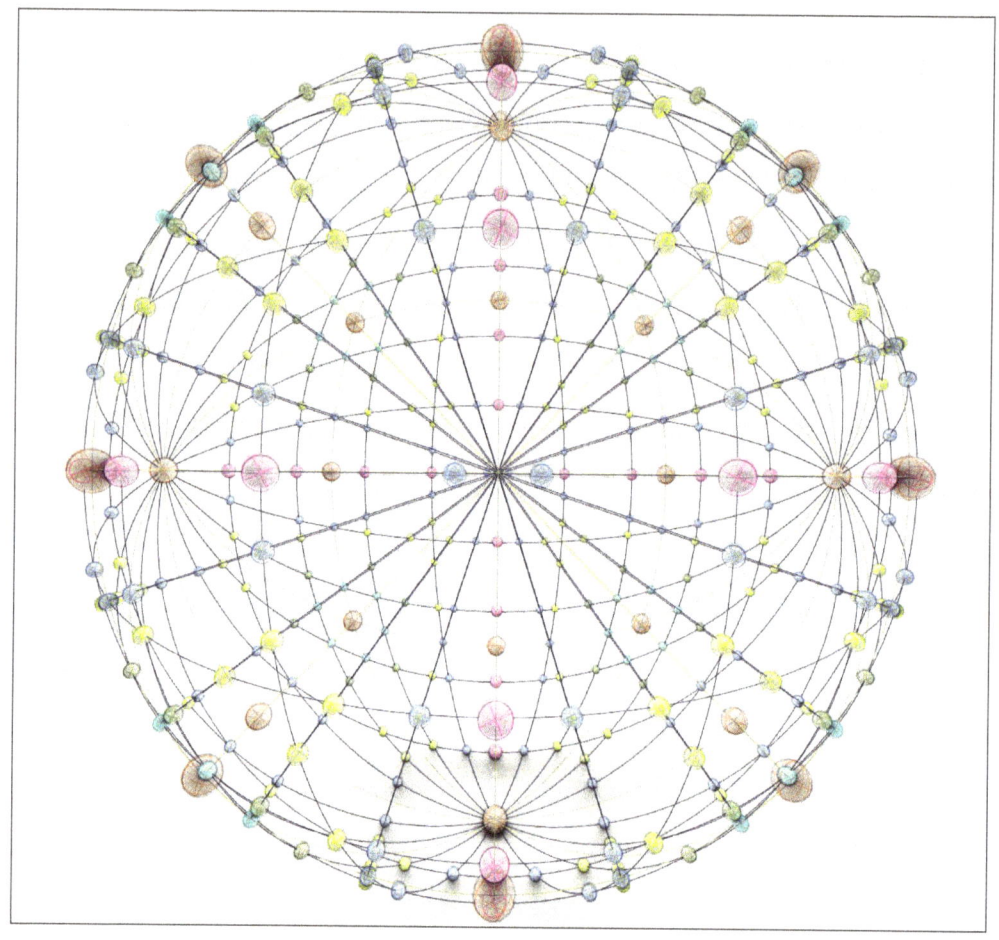

A crystallographic grid with axes of symmetry and a color coded coordinated system

Crystallographic tectonic exploration

Deconstructed box tectonics an exploded axonometric drawing

Deconstructed box tectonics an exploded axonometric drawing forming matrices in space

The many formats and types of abstract geometry generate potentials for the aggregation of space as a multimedia capacity for the architectural designer. The variant systems at play lend qualities to the organization and assemblies of space. The degree to which we can derive a sophisticated geometric system is our ability to bring harmony to our complex realities. As we invent novel constructs for the assembly of geometry we become more conscious of our ability to assimilate more expansive aspects of our reality. As a technique, geometric operations consolidate the mysterious aspect of puzzling configurations. From crystallographic systems to urban grids and organic tissues, morphology defines an operative reality, potent with information, energy, and embedded qualities of compositional integrity. The exploration into geometry is a discovery process that invents new methods of constructing our surrounding world and objects. The role of our built-environments to sustain the lives of the biodiversity through their infrastructure in the shifting and moving dynamics of subject matter is accomplished by geometrical compositions. Our multi-layered grids and infrastructures are a geometrical construct, a matrix of components. There is a multi-scalar nature to the geometry of terrestrial architecture. Propagating from single cell components that multiply and diffuse becoming vast grid networks of distributed energy. Our knowledge of geometry and its compositional capacities offer us our aggregation potential. With the many dynamics of the built-environment and its embedded complexity, residual ways of life, geometry and morphology are our tools of integration. The needs of all can be met with sophisticated design thinking as we allocate with coordinate precision the distributions, absorptions and storage of our multidimensional energy conduits.

CHAPTER 14

Morphogenesis, Epigenesis, and the Continual Meta Morphologies of Architectural Genealogies

The living status of land guarantees a residual infrastructure. In order to accommodate the biodiversity in a domain of land, residual organizations are assembled towards a management of resources that occur through the synthesis of material assembly with the organizational capacities of geometry. The residual organizations are each enacted with an architecture that instills their solutions with responsive specificity. In accumulation, there is a formation or a residual morphological act of architecture. With rule sets, in chain reaction, the acts of architecture discern an incremental and exponential formational structure of resource allocation, amassing the city.

A resource available through a designed infrastructure is a redistributed conglomerate force, often approached through the masterplan. The recursive cycle of generating the city includes the management of such resource conglomerations. The management ensures a productive recursive cycle of nodal points that mitigate reconstruction or reorganization of material in both micro or macro activities. The nodes that discern such organizational or constructional capacities are located within constituent mediums that are scalar. The nodes could either serve as sources of building material or locations of platform management.

Platform management systems can be considered as conscious techniques of material assembly and allocation, an act that in turn should have the potential to evade the accidental and enter a compositional accuracy in accumulation—through a conscious awareness of unfolding actions provided by the feedback loop of cybernetic analytic systems. Rules and axioms are formal systems that enable one to intuit directions and elaborations in a continual expression in validity with coherency and growth consequences. Available techniques of informational management are actions attributed with a vocabulary that establish location within a sequence and a residual behavior or action with such contextual relativity. The layering of architectural methodologies can continuously interchange pressures as a symphonic layering of music or a sequential proof of mathematics in the many contexts that follow. Themes and uses of architecture function in programmatic duals, triples, and quadruples—in a multivalence that morphological translations reconcile. A continual dialectic of embedded code and rules is in a morphological system and the associative adjacency of contextual systems.

Morphological transformations of the city are repercussions residual to the intentions of users with given incentives to invest time and resources into a compliance with the contextual acts. Directly, an analytic extrapolation of component architectures in infrastructure serves as an understanding of circumstance in tangential casualties that are given by the lexicological structures of a system.

What we can infer or automatically prove through the methods of incremental consideration, through valence or multivalence as a functional context is that the instrumentalization of a single theme or component may be proliferated, supported, activated, and manipulated into a continual system of (melodic/harmonic) accumulations. Material proportions and multi-valences of performance are accessed in a simultaneity of assemblies that are utilized as continual cohabitations with the living beings of an ecology. Mass energy distributions, allocations, and circulations are apparent in global scales as accessible modalities of constituent forces. The awareness of contextual accuracy and potential proliferation of an infrastructural act discerns its capacity for instilling a continual multi-valence of accessible stimulations for following acts.

In the simultaneous sequential contrasting and potentially harmonious initiations of private interests upon the built environment, phasing structures amass, accumulate, and assemble in extended sequential layerings of project initiatives. The situations are spatially pertinent or in other circumstances contextualized by their platform accessibility through modes of multimedia that an architectural morphology may obtain. The multimedia existence of an architectural morphology is the versatility of its accessibility and manipulability. Morphological epigenesis as a manipulation of material and geometry begins in a temporal precision of relative action. The virtual presence of the architectural morphology can be simulated in the platform as many

potential design solutions.

The continual metamorphology of landscapes and buildings in an urban typology as a gradual evolution into a rearrangement of material with given infrastructures is connected the coordination of the public body—to organize available resources and deploy a coordinated extended measure as a consciousness towards a natural landscape, social scape—a digital/mechanical arrangement of production and consumption models.

Waterways, roads, and railways are constructions of a communicative network system, an accelerative model of exchange. Each distributive model affords a direct compatibility with a type of resource that has geographical specificity, such as a basin, river, or existing hydrology system. The meta-morphologies of the cities of global scale are inextricably interwoven with distributed protocols that weave the morphological progress into cultural acts relating to natural resources. From platforms, the media translations of incremental morphological acts of architecture are made visible in a perspective that provides the opportunity for a meta-morphology. In perspective of market growth or resource allocation, the locations of activity that are revealed through mediums of recording reveal patterned behaviors that enable a cross referencing of cartographic representations. Revealed through the cartographies is a meta-morphology or transformation of increased scale and duration.

The multiple agents contributing to cartographic depictions of environmental space have tendencies and lineages of technique that are heuristically sensible. Conscious awareness and intentions are not individual submissions or solipsistic interests when the built environment in duration and scale serves the interest of multiple life forms and has a myriad of consequences for collective living. Conscious awareness and intentions of constituent agents with respective capacities and abilities are transmitted among translations that the lexicological structures of cartographies afford as interdisciplinary action takes place. Domains of intelligence intersect domains of physical topographical and urban boundaries among a sovereignty. The context of the contributors to the informational platforms of built environments is a collection of representations in communication and projection, to interface the techniques of architectural morphologies that are an ultimate translation of cultural intelligence into material architecture.

The collective intelligence or conscious awareness of a populous to equip itself with typologies of spatial intelligences defines an operative protocol with habitualized referencing and embedded intention. The collective action or incremental intentions are supported by a heuristic library of morphological tactics that have been trialed in academic settings or in independent explorations to find deployment via relevancy or by other preferences of sovereign abilities. Or in other less thoughtful environments, the morphological acts of architecture sprawl without academic trial or consideration, excluding any possibility of integrating the architecture into a shared media. The potential of sovereign abilities is enabled by the heuristic library of morphological tactics and the scope for envisioning provided by platform architectures that integrate multiple models and maps. Morphological acts are contextualized on the global platform in physiological performance and scholastic relativity, only possible by the driven formation provided by agencies as groups of people.

The simultaneity of these accurate descriptions guarantees that the multimedia modes of definition arrive as projections of worlds as perspectival images, cartographic images, infrastructural models, and sociological models—from given variables that re-articulate reality through a construct. The construct as a method of understanding is an occupied mental space, a location of cerebral tendency that is visited through the cartographic image as a constructed understanding. The codex of a city and its mapped uses of natural resources guarantees a behavior pattern that engages material substrate protocols that result as cultural phenomena with commanding momentum. Narrative and constructs are able to be derived by interlinked systems; yet so presently pertinent are the existing morphologies of nature that also carry a commanding influence.

The geometrical representations as exchanges and codependent dynamics ensure that there are informational existences that are represented and accessed by charts and legends of statistical

understandings. The social understanding of these dynamics has a cultural repercussion of societal hierarchies and substrates designated towards a pattern relationship with the consumption processing reorganization and awareness of the codependencies. Their uses ensure a support of various industries and labor sequences around the further management of the codependency. With methodological architectural transformations that have morphological effect, sovereign subjectivity gives accessibility to the exchange productions, consumptions, and transportive models that are included within a landscape. These methods of extraction are embedded in infrastructural architectures, giving the ability of interrelation of infrastructures to their environment.

The evolution of the planet is dependent upon our peaceful coexistence that will enable new forms of creations that are planetary and environmental paradigms of architecture that are environmental, ecological, and planetary in scale. As we become familiar with geospatial cartographies and algorithmic tool sets our scopes of design influence increases to encompass environments. As we become more conscious, peacefully coexisting our ability to communicate and coordinate geometric operations within the city become co-functional. Infrastructurally the needs of many can be met with well coordinated design and the inspired intentions of a collective consciousness can be expressed in a collective terrestrial architecture.

The following images I've constructed with AI will be the concluding note. These are abstract geometries as prototypical cartographies with the potential to be Terrestrial Architectures.

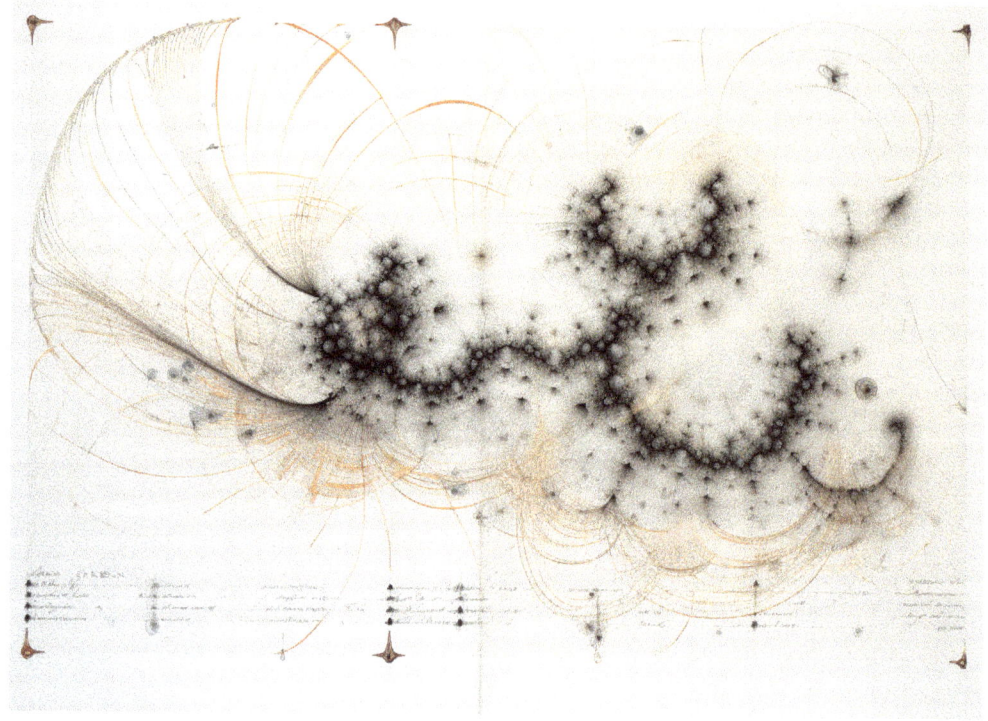

An organic radial grid structure which begins to resemble the cellular diffusion of an organic city.

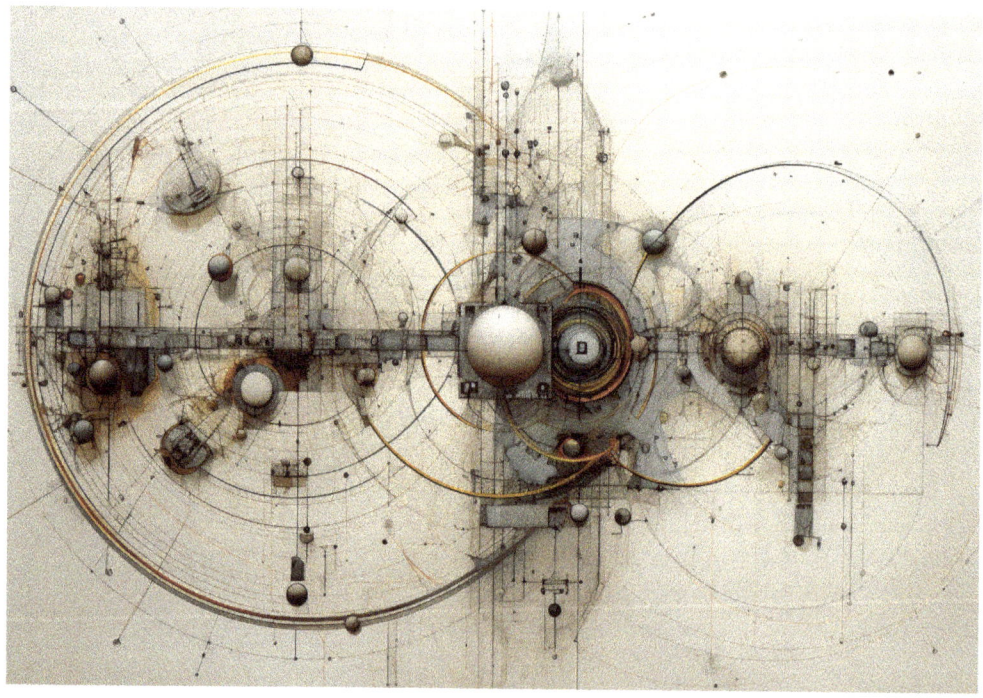

A more mechanical or celestial model of a radial grid system yet still hinting towards the structure of a city.

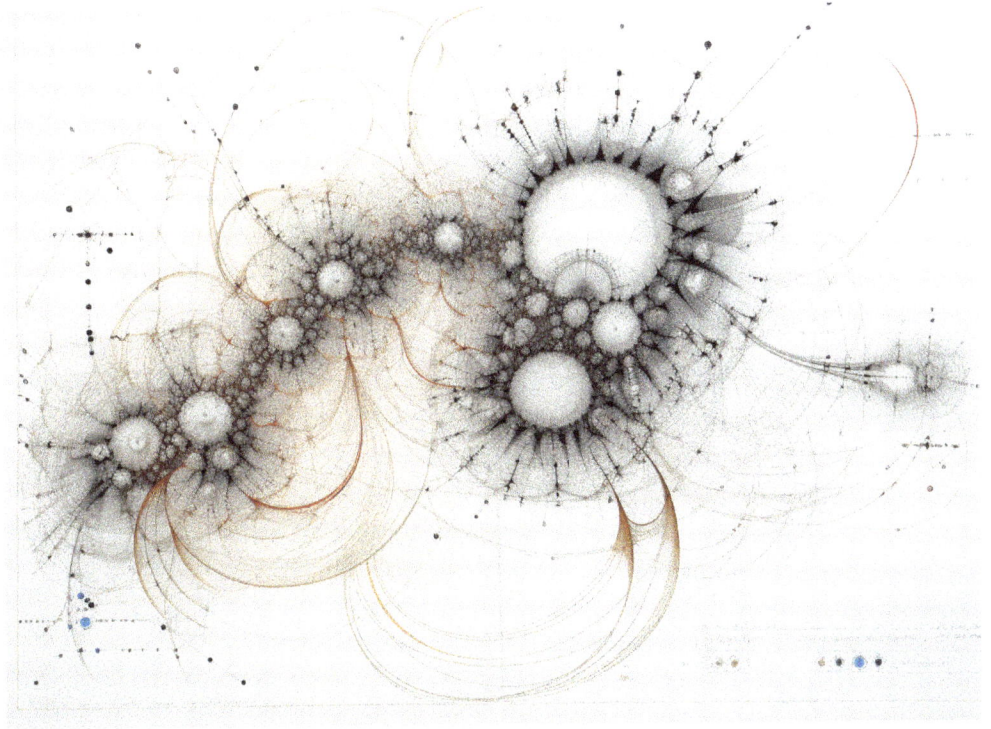

Radial grid structures fused with organic membrane tissues creating what seems to be an aerial view of a habitation of sorts.

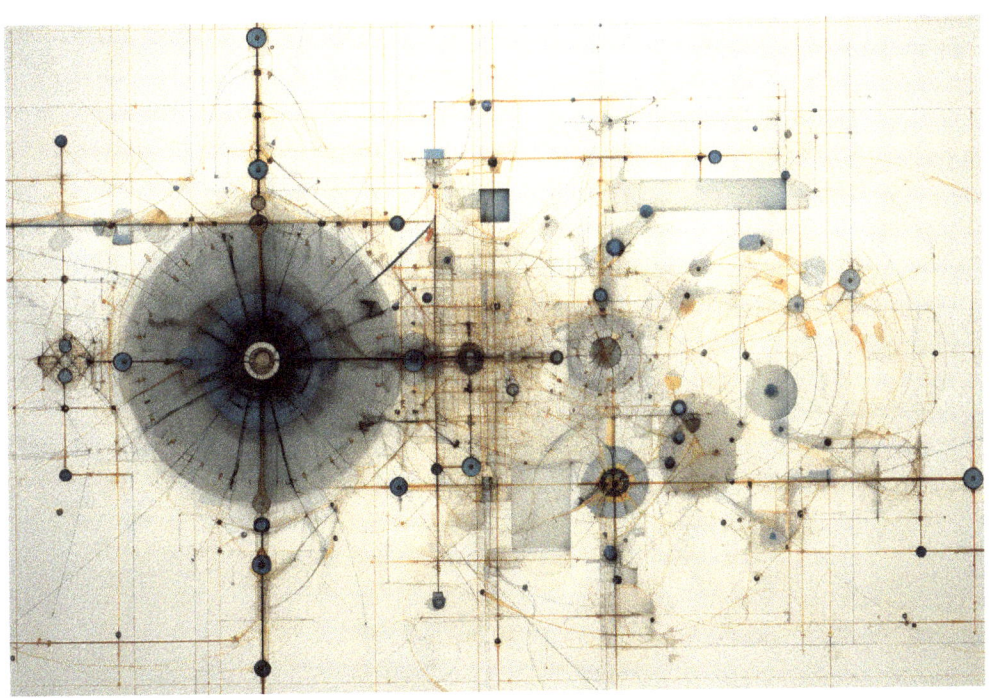

Radial grid structures with axes forming an aggregation of clusters - another abstract city schematic.

Projected along a main axis, a series of multi-scalar circles creates a radial grid structure.

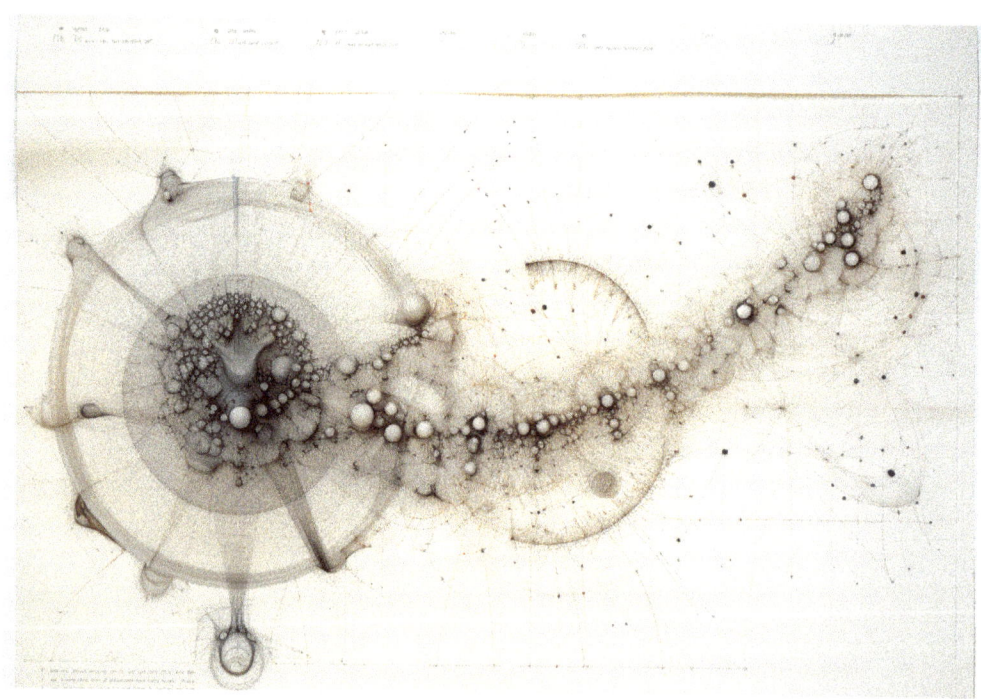

A radial grid structure is fused with organic tissue membrane structures creating an arpeggiation across the space of the drawing.

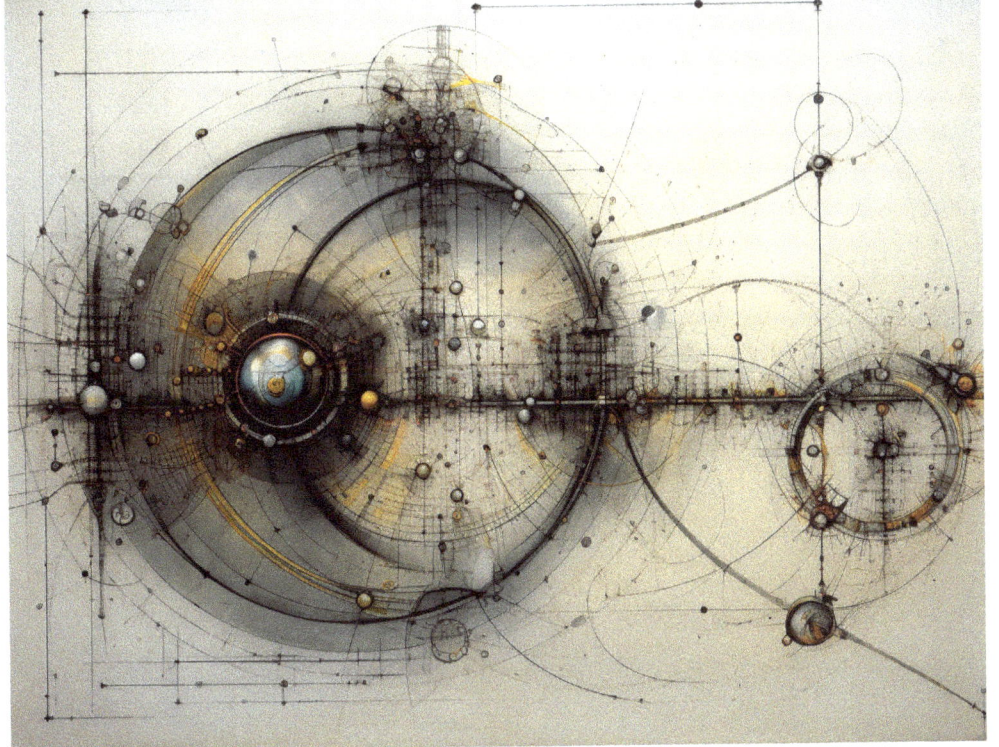

Radial grid structure with many specific details contained within an overall structure.

The radial grids are concentric offset from each other and embedded within themselves as a recursive grid.

More organic radial systems combined with a series of extruded tower members forming a collage of morphological systems.

Landsat imagery combined with mathematical geometries from Mandelbulb in AI image generating software Midjourney.

Seeking to tectonically describe a city with a radial motherboard aesthetic.

Radial machine components reference motherboards and quantum computers.

Peripheral radial structures are part of the formulation as the geometric system is expansive.

Visualization in Unreal Engine

An abstract geometrical grid encompasses the globe revealing a potential for an architecture of the Earth.

Acknowledgements

The writing and research happened with the influential guidance of many mentors, including Andrew Witt, Greg Lynn, Casey Rehm, Benjamin Bratton, Patrik Schumacher, Ben Van Berkel, Graham Harman, Timothy Morton, John Enright, Hernan Diaz Alonso, John May, Jorge Silvetti, Jennifer Bonner, Peter Zellner, Jake Matatyaou, David Ruy, Todd Gannon, Tom Wiscombe, Eve Blau, Dwayne Oyler, Sotoru Sugihara, Ung-Joo Scott Lee from Morphosis Architects, Jordan Kanter when we worked together at MAD Architects in Beijing, my parents Michael Rendler AIA Director of e7 Architecture Studio and Professor Marcela Oliva AIA Educator of the Year 2022, my Karate Shihan Patrick Fard who trained me to discipline my mind from a young age, my high school English teacher Beth Jones, who first introduced me to sublime teachings through the novel Frankenstein, Jose Negrete and Alex Abarca. Thank you to my mentors.

Thank you to my heroes: Zaha Hadid, Peter Eisenman, Lebbeus Woods, Thom Mayne, Kobe Bryant, Lionel Messi, Pope Francis, Leonardo Da Vinci, Galileo, Martin Luther King, Gaudi, and of course my family friend, Glen Howard Small.

A special thank you to the National Aeronautics and Space Administration (NASA) and Vera with Mars City Design who presented me with the Mars City Design Urban Category 1st Prize Award which encouraged me to think on the terrestrial scale.

Thank you to my favorite movie director Christopher Nolan.

Thank you, George Lucas.

Thank you, Hayao Miyazaki.

Thank you, J.R.R. Tolkien.

Thank you to the creators of *Avatar: The Last Air Bender*.

I am also grateful to the wonderful colleagues who inspired me with good company and dedicated passion for their studies: Aakash Shah, Connor Gravelle, Jordan Hartman, Morgan Garrard, Kevin Arango, Miwa Espinoza, Karim Saleh, Luciano Menghini Javier Benavidas, Matt Pugh,

Aimilios Devlantes Lo, Savina Hawkins, Alex Spenzanris Alkiviadis Pyliodis, Evangelos Fokialis, Debbie Garcia, Benjamin Nelson Pennel, Benjamin Pollak, Andrew Smith, Isabella De Sousa, Cassidy Viser, Kevin Shah, Adrian Wong, Ryan Baits, Anna Goga, Meric Arslanoglu, Danielle Rose, Daniel Berdichevski, Neeraj Mahajan, Juan David Grisales, Mariel Collard, my Harvard dorm kin in Childe Hall: Sol Greene Eames, Josh Dean, Suhyun Kim, Tasos Giannakoupoulos, Stephane Tcho, Dylan Rupar.

Thank you to John Clippinger and David Thomson for the Gaia Windhover Conference experience and all who were there.

Thank you to my West Coast friends Kyle Weissman, Nadav Ben Dayan, David Shaby, Chris Madrid, and Sammy Nunan.

Thank you to my friends—Nona, Kostas, Shuei, Kang, Tul, Melody—in Beijing at MAD who were a kind presence when I dived deeper into the studies of algorithms.

Thank you to MAD, Morphosis Architects, and Gensler for the learning opportunity.

Thank you to my childhood friend Willy Rudy and to all of the Rudy Cody Lynch Families

Thank you to my family in Mexico and all my aunts and uncles.

Thank you two wonderful sisters Vineta and Mariana whose hearts shine in my life so radiantly. Thank you to the consistent support of my brother Billy Marrone

Thank you to my dear Robin and the Woerner Family

Thank you to all my family and every thinker in the bibliography who has contributed towards this vision.

Thank you to Jill L. Ferguson, the editor of this book, without whom this book wouldn't be possible.

I am so grateful to live, think, and create with all of you.

Bibliography
(in order of the appearance within the chapters):

Pope Francis. *Laudato Si': On Care for Our Common Home* [Encyclical], 2015.

Lovelock, James. *GAIA: A New Look at Life on Earth.* Oxford: Oxford University Press, 2016.

Small, Glen. "Biomorphic Biosphere 2." Small at Large, 2013. http://www.smallatlarge.com.

Verndasky, Vladmir I. *The Biosphere.* Krakow: Copernicus Press, 1995

Verndasky, Vladimir I. *Geochemistry and the Biosphere: Essays.* Santa Fe, NM: Synergetic Press, 2018.

Pope Leo XIII. *Summa Theologiae.* [Encyclical], 1893.

Harman, Graham. *Object-Oriented Ontology: A New Theory of Everything.* New York: Pelican Books, 2018.

Latour, Bruno. *Facing Gaia: Eight Lectures on the New Climatic Regime.* Cambridge: Polity Press, 2017.

Sheldrake, Rupert. *Morphic Resonance: The Nature of Formative Causation.* Rochester: VT: Park Street Press, 2009.

Lynn, Greg. *Animate Form.* Princeton, NJ: Princeton Architectural Press, 1999.

Hoftsadter, Douglas. *Gödel, Escher, Bach: An Eternal Golden Braid.* New York: Basic Books, 1979.

Weiner, Norbert. *Cybernetics: Or the Control and Communication in the Animal and the Machine.* Eastford, CT: Martino Fine Books, 2013.

Zhang, Yongjun, et al. "Fully Automatic Generation of Geoinformation Products with Chinese zy-3 Satellite Imagery." *The Photogrammetric Record.* 16 Dec. 2014. https://doi.org/10.1111/phor.12078.

Ramachandran, Bhaskar, et al (ed.). *Land Remote Sensing and Global Environmental Change: NASA's Observing System and the Science of ASTER and MODIS.* Springer, 2010.

Bratton, Benjamin. *The Stack: On Software and Sovereignty.* Cambridge, MA: MIT Press, 2016.

Hardt, Michael. *Gilles Deleuze: An Apprentice in Philosophy.* Minneapolis, MN: University of Minnesota Press, 1993.

Rheinberger, Hans-Jorg. "Gaston Bachlard and the Notion of Phenomenotechnique." *Perspectives in Science*, Vol. 13, 313-328, 2005. https://doi.org/10.1162/106361405774288026

Schapp, Jeffrey T. "Knowledge Design: Incubating New Knowledge Forms/Genres/Spaces in the Laboratory of the Digital Humanities." Keynote delivered at the Herrenhausen Conference, 2013.

Harvey, David. "A 'Grid' of Spatial Practices", from 'Flexible Accumulation through Urbanization Reflections on "Post-Modernism" in the American City.' *Perspecta* 26. Hans Baldauf, Baker Goodwin, Amy Reichert, eds. New Haven, CT: Yale School of Architecture, 1990.

Giedion, Siegfried. *Mechanization Takes Command: A Contribution to Anonymous History.* Minneapolis, MN: University of Minnesota Press, 2014.

201

Biography

Jack Oliva-Rendler is exploring work processes in Digital 3D modeling, Geospatial Information Systems (GIS), and algorithmic modeling methods for multi-scaler ecological solutions, planetary, bio-regional, infrastructural, local and community based scales of work. He attended the Southern California Institute of Architecture, where he developed skills in metaphysical and computational philosophy and also design technology to win an Undergraduate Thesis Prize. He also received a Masters Degree in Architecture from the Harvard Graduate school of Design. During his graduate studies he received an award from Mars City Design, in the Urban Design Category, in the Mars City Design Competition, with panelists and judges from NASA and Dassault Systemes. Jack's thesis projects at both institutions proposed radical data infrastructures for Earth Observation Science and environmental analysis. His design specialities have evolved to include fractal formulations and mathematical simulations in Mandelbulb 3D as ways to simulate organic and crystallographic fractal systems with a multitude of applications. His algorithmic fractal modeling techniques have been launched as an in-depth online education course on the Futurly Platform.

www.ingramcontent.com/pod-product-compliance
Lightning Source LLC
Chambersburg PA
CBHW041535120626
46551CB00019B/2700